DATE DUE

DE 2 6 '00			
OC 16 '01			
NO 26 02			
FE 12 '04			
NO 2 7 '04			
MY 1 5 08			

DEMCO 38-296

Tasting and Smelling

Handbook of Perception and Cognition

2nd Edition

Series Editors
Edward C. Carterette
and **Morton P. Friedman**

Tasting and Smelling

Edited by

Gary K. Beauchamp

Monell Chemical Senses Center
Philadelphia, Pennsylvania

Linda Bartoshuk

Yale University School of Medicine
Department of Surgery
New Haven, Connecticut

Academic Press

San Diego   London   Boston
New York   Sydney   Tokyo   Toronto

/

This book is printed on acid-free paper ∞

Academic Press
a division of Harcourt Brace & Company
525 B Street, Suite 1900, San Diego, California 92101-4495, USA
http://www.apnet.com

Academic Press Limited
24-28 Oval Road, London NW1 7DX, UK
http://www.hbuk.co.uk/ap/

Library of Congress Cataloging-in-Publication Data

Tasting and smelling / edited by Gary K. Beauchamp, Linda Bartoshuk
        p.   cm. -- (Handbook of perception and cognition, 2nd edition)
    Includes bibliographical references and index.
    ISBN 0-12-161958-3 (alk. paper)
    1. Taste. 2. Smell. I. Beauchamp, Gary K.   II. Bartoshuk,
    Linda.   III. Series: Handbook of perception and cognition (2nd ed.)
    QP456.T372   1997
    612.8'7--dc21   97-22352
                                                      CIP

PRINTED IN THE UNITED STATES OF AMERICA
97   98   99   00   01   02   EB   9   8   7   6   5   4   3   2   1

Contents

3   *Psychophysics of Taste*

Bruce P. Halpern

4   *Olfactory Psychophysics*

Harry T. Lawless

5 *Clinical Disorders of Smell and Taste*

Beverly J. Cowart, I. M. Young, Roy S. Feldman, and Louis D. Lowry

Contributors

Numbers in parentheses indicate the pages on which the authors' contributions begin.

Gary K. Beauchamp (199)
Monell Chemical Senses Center
Philadelphia, Pennsylvania 19104

Joseph G. Brand (1)
Monell Chemical Senses Center
Philadelphia, Pennsylvania 19104

Veterans Affairs Medical Center
Philadelphia, Pennsylvania 19104

Beverly J. Cowart (175)
Monell Chemical Senses Center
Philadelphia, Pennsylvania 19104

Department of Otolaryngology
Head and Neck Surgery
Thomas Jefferson University
Philadelphia, Pennsylvania 19107

Roy S. Feldman (175)
Veterans Affairs Medical Center
Philadelphia, Pennsylvania 19104

Bruce P. Halpern (77)
Department of Psychology
and Section of Neurobiology and Behavior
Cornell University
Ithaca, New York 14853

Harry T. Lawless (125)
Department of Food Science

New York College of Agriculture and
  Life Sciences
Cornell University
Ithaca, New York 14853

Louis D. Lowry (175)
Department of Otolaryngology
Head and Neck Surgery
Thomas Jefferson University
Philadelphia, Pennsylvania 19107

Julie A. Mennella (199)
Monell Chemical Senses Center
Philadelphia, Pennsylvania 19104

David V. Smith (25)
Department of Anatomy and Neurobiology
University of Maryland School of Medicine
Baltimore, Maryland 21201

Mark B. Vogt (25)
Department of Anatomy and Neurobiology
University of Maryland School of Medicine
Baltimore, Maryland 21201

I. M. Young (175)
Department of Otolaryngology
Head and Neck Surgery
Thomas Jefferson University
Philadelphia, Pennsylvania 19107

Preface

This book consists of six chapters covering a variety of topics on taste and smell research. It is not intended to be a complete overview of research in the chemical senses. The authors of these essays were encouraged to emphasize their own perspectives on important issues in the field. They were particularly asked to address unanswered questions and neglected research topics. Consequently, each of the chapters provides a point of view on an important and often controversial research area in the chemical senses. The editors also chose not to include chapters in several areas that have been thoroughly and frequently reviewed. For example, olfactory transduction and CNS processing have received considerable attention elsewhere and thus are not treated here.

Each of the first three chapters is concerned with the sense of taste. As befits this research area, the question of whether there are "primary" or "basic" tastes (e.g., whether taste experience can be classified into a small number of categories, namely, sweet, sour, salty, bitter, and perhaps a few others) is implicitly or explicitly a central issue in all three chapters. In his overview of the biophysics of taste, Brand argues in Chapter 1 that the sense of taste can be conveniently divided into discrete categories, most likely four or five, and he discusses the transduction mechanisms underlying these categories. Within each of these categories, he shows that multiple receptor and transduction mechanisms exist. The remarkable progress in unraveling these molecular and cellular processes forms the bulk of this chapter.

Smith and Vogt, in Chapter 2, on neural codes and integrative processing of taste, argue that the goal of the neurophysiologist is to trace the pathways and elucidate the mechanisms of information processing throughout the central nervous system. They point out, however, that because individual taste fibers are often responsive to stimuli that elicit more than one of these categorical experiences (e.g., sucrose and NaCl; sweet and salty), these categories cannot arise through a strict straight-line mechanism where each fiber carries information on only a single taste quality. Moreover, taste nerve fibers also are influenced by tactile and other nontaste stimuli, further confusing the issue. Nevertheless, Smith and Vogt marshal an impressive array of evidence indicating that while no individual taste fiber is exclusively responsive to a single taste quality, the fibers do have a rough specificity. From this complexity, Smith and Vogt delineate the reasons that there is still sentiment for the idea that taste coding is similar to coding in color vision. They argue that, in taste, activity of one fiber type is insufficient to discriminate between stimuli of different taste qualities; that is, salt-best fibers do not, by themselves, signal saltiness.

Halpern, in his treatment of the psychophysics of taste, in Chapter 3, focuses on a perceptual phenomenon, the nature of taste mixtures. He provides a sophisticated account of the complexity inherent in the old debate over whether taste mixtures are analytic (i.e., the components are individually perceived) or synthetic (i.e., the components lose their identities and a new quality emerges). This old debate remains important because of the argument that it is linked to the coding debate (e.g., there is a presumption that labeled-lines would result in analytic mixing while quality coding dependent on multiple fibers would result in synthetic mixing). Halpern goes on to argue that the general acceptance by researchers of a small number of taste qualities is premature and dangerously influences the kinds of experiments investigators undertake. Instead, he draws the reader's attention to the taste complexities of real foods and to studies by investigators who argue against what he calls the basic taste theory.

In considering the utility of basic taste theory, Brand and Smith and Vogt note an interesting taste phenomenon: certain chemical compounds appear to be able to specifically eliminate one or more of the basic tastes without substantially altering other taste sensations. For humans, one dramatic instance of this is the effect of lactisole (the sodium salt of 2-[4-methoxyphenoxy]-propanoic acid) on sweet taste. It appears that this compound blocks sweetness of all substances (with perhaps a very few exceptions). This observation may have profound implications for understanding the molecular mechanisms for taste transduction and for theoretical considerations concerning the existence of a small set of taste experiences (not compounds) out of which all others are constructed. If we were able to identify similar specific blockers of bitter, salty, sour, and perhaps umami taste qualities, would every sapid substance be thereby rendered tasteless? If for no other reason than testing this hypothesis, a search for such blockers is of great theoretical as well as obviously practical interest.

Beyond the issue of basic tastes or taste primaries, Brand's chapter provides a fine overview of the fascinating details of taste transduction and provides hints as to how knowledge here—sometimes ignored by those interested in behavior—should have a profound impact on our understanding of taste perception. Smith and Vogt emphasize the role of taste as a hedonic system and one that impacts ingestive behavior and physiology. That taste is a sensory system with but a single major function having to do with food acceptance and utilization is often neglected. The importance of understanding how quality information is integrated with information on acceptance, rejection, and utilization is emphasized in this chapter. Halpern, in his own research as well as in his chapter, highlights the temporal aspects of taste. Taste perception changes over time, which adds complexity (which investigators often wish to ignore) to both intensity and quality evaluations. This also opens the intriguing question, touched on briefly by Halpern, of whether one can have intensity without quality.

Chapters 5 and 6 also deal in part with taste, although both focus more on olfaction. In Chapter 5, Cowart and colleagues draw needed attention to historical observations in their syntheses of published research and their own new data on chemosensory disorders. They note that the modern finding using rigorous psychophysical procedures of many more olfactory than taste disorders in the general population was noted as long ago as 1884 by Mackenzie, who referred to even earlier work. Nevertheless, this chapter shows the considerable progress that has been made in diagnoses and in understanding the prognoses for olfactory and taste disorders during the past 10 or so years.

In their consideration of taste disorders, Cowart *et al.* make the interesting and important observation that each of the so-called basic qualities can be independently affected. This observation may be interpreted to lend further support to the validity of this kind of categorization. However, in discussing olfactory disorders, they note that loss seems to be general rather than specific. That is, they refer to no cases of specific loss of a "quality" of odor (with the exception of specific anosmia)—no one reports loss of sensitivity to citrus odors, for example. In fact, modern clinical research has shown very high correlation in patient populations between odor thresholds and odor identification, suggesting that losses, when they occur, are of a very general nature.

This observation is consistent with many of the points that Lawless makes in Chapter 4, on olfactory psychophysics. In this extensive overview of many aspects of human olfactory perception, he returns several times to the issue of odor classification: Are there a small number of "primary" odors out of which the complex real odor world we deal with is made? The answer, according to Lawless, is a resounding no. In particular, effects of odor mixing, emergent properties in complex mixtures, the unitary and unique nature of some mixtures, and the (presumed) ability of experts to identify and categorize a vast number of unique odors and odor qualities all lead to the conclusion that many different quality terms (and hence sensory percepts?) will be needed to adequately categorize odor quality. Indeed, the very categories of one's perception of odors may depend on experience. The perfumer uses different odor categories than the wine expert based on their differing domains. However, in an interesting exercise, Lawless does demonstrate that perhaps there is more agreement about our odor categories than most people have recognized.

This raises the question whether similarities in quality judgments represent deep cultural similarities or perhaps physiologically based patterns of similarity (p. 163). One obvious potential source of a physiological basis for universal similarity is the way in which olfactory receptor families may be organized. It will be intriguing to see, as we learn more about receptor sensitivities, whether classification schemes based on psychophysics and biophysics are isomorphic. There is danger in necessarily expecting such a congruence, however, which is alluded to by Lawless: the major role in all aspects of olfactory perception played by learning and cognition. It may be naive to assume that knowledge of receptor mechanisms will tell us much about odor classification at the psychophysical level.

The issue of the importance of learning, particularly with regard to olfaction, is taken up in the last chapter on development by Mennella and Beauchamp. Following a brief overview of early development of the taste and olfactory systems in human fetuses and infants, this chapter focuses on the flavor (primarily retronasal olfactory component) world of the developing human infant. These authors suggest that exposure to flavor in amniotic fluid and human milk may contribute to later preferences for such flavors.

In differing ways, all the authors in this volume are confronted with a fundamental and, to date, intractable problem: while as scientists we like to analyze the simplest of systems in order to better manipulate them and obtain interpretable data, as smelling and eating individuals we also know that in the real world tastes and smells are almost always very complex mixtures. It is unclear whether the principles obtained from studies on simple systems will be sufficient to explain the real world data or whether new phenomena emerge with complexity. Without a much better understanding of how the central nervous system processes information, we cannot answer these questions. Yet it is important to keep in mind, as do all the authors here, that what we must ultimately explain are the biological psychological bases for the experiences Halpern describes in the opening section of his chapter, the feats that Lawless attributes to perfumers and the like, the profoundly unpleasant experiences Cowart et al. describe for patients with dysosmia and dysgeusia, and even the presumed complexity of the flavor world of the human fetus.

We have enjoyed working with the authors on their chapters over a rather longer period than we or they anticipated. We trust the readers will enjoy these chapters as well.

Gary K. Beauchamp and Linda Bartoshuk

CHAPTER 1

Biophysics of Taste

Joseph G. Brand

I. INTRODUCTION

Taste transduction is initiated when a sapid stimulus interacts with a specific receptor entity of a taste receptor cell. Taste receptor cells are neuroepithelial cells that are usually clustered into multicellular arrays called taste buds. In mammals these buds are situated in special papillae or regions of the oral cavity. Taste cells form synapses with innervating sensory neurons, and are in contact with each other through numerous gap junctions (for reviews, see Kinnamon, 1987; Roper, 1989). The taste bud, being an epithelial structure, is functionally divisible into two parts: the apical region and the basal region. The apical region faces the oral cavity and is the site of interaction of most stimuli with taste receptor cells. The apical region and the lower basal region are separated by tight junctions that, as they do in all epithelia, effectively limit the diffusion of many compounds from the mucosal to the serosal space. This barrier insulates the intracellular space around the basal region from most substances that enter the oral cavity. This diffusion barrier has important implications for taste transduction, because it limits the area of interaction of stimuli with the receptor cell membrane and acts as a selective filter for many ions, thereby directly affecting the magnitude of some tastes, as is the case for salty taste transduction (see later). Within the taste bud, the stem cells that give rise to taste receptor cells are called basal cells. Under certain circumstances, these cells may play a role in the transduction process (Roper, 1992). The general features of the taste bud are outlined in Figure 1.

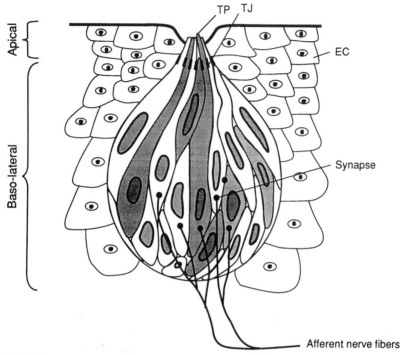

FIGURE 1   Schematic drawing of a typical mammalian taste bud. The taste bud is composed of approximately 75 to 150 specialized epithelial cells, some of which make synaptic contact with innervating sensory nerve fibers. Only a few of the cells of the taste bud are exposed to the oral environment at the taste pore (TP) at any given time. Because of the presence of tight junctions (TJ) just below the TP, the taste bud, like other epithelial cells (EC), functions as a partial barrier to the diffusion of large molecules and large ions. These tight junctions effectively divide the taste bud into a small apical region and a larger basal lateral region, polarizing the cells with respect to their extracellular environment. The taste receptor cells presumably insert receptors for taste stimuli into the plasma membrane, with transductive elements located in the membrane and the cytoplasm. A sufficiently large transductive wave can initiate secretion of neurotransmitter from the taste receptor cell to the innervating sensory nerve at the synapse. There is evidence also for synaptic contact between apparent taste receptor cells and basal cells within the taste bud (Reutter, 1978; Roper, 1992), with the basal cells then showing synaptic contact with sensory nerve fibers. The function of this intervening cell is not understood, but it is possible that it may act as an interneuron does in the visual system, allowing a hyperpolarizing response of the receptor cell to be transformed into an excitatory response at the neural level.

The sense of taste is most conveniently divided into a number of discrete modalities. Although there has been some disagreement as to the number of these and their precise description, it is generally assumed that this sense transduces at least four or five basic modalities. These are salty, sour, sweet, bitter, and umami. This last modality, still a putative one, is described by a Japanese word, which roughly translates into savory or delicious. This taste is defined by the taste of its prototypical stimulus, monosodium-L-glutamate. Whether umami is a unique modality, separate from the other four, or is the result of a mixture of sensations of two or more of the other modalities, is still a matter of some debate. There is growing evidence, however, that at least some of the transduction sequences for this fifth taste are unique.

It has become increasingly clear over the past decade that unique receptor mechanisms exist for each of these modalities. (For recent reviews, see Akabas, 1990; Avenet & Kinnamon, 1991; Brand & Bryant, 1994; Brand & Feigin, 1996; Corey & Roper, 1992; Gilbertson, 1993; Kinnamon & Cummings, 1992; Kinnamon & Margolskee, 1996; Kurihara, 1990; Lindemann, 1996; Margolskee, 1993; Naim, 1993; Tonosaki, 1990). These receptor processes can be roughly divided into three types: (1) those making use of a presumed receptor protein that transduces its interaction with a stimulus to change the levels of various intracellular signaling compounds (i.e., a second messenger signaling system); (2) those making use of existing plasma membrane-associated ion channels in taste cells, whereby the stimulus traverses the membrane through these channels or modulates the activity of these channels; and (3) those making use of a stimulus-gated ion channel-type receptor complex, whereby a taste stimulus binds to a receptor site on a receptor–ion channel complex and alters the flux of ions through the ion channel. Each of these transduction mechanisms ultimately alters the ion balance within the receptor cell. This change in ion activity within the cell is usually accompanied by, or caused by, changes in ion flux across the plasma membrane of the taste receptor cell.

Taste receptor cells are innervated by sensory nerve fibers from one of three cranial nerves: VII, IX, or X. In order to report the presence of a taste stimulus to the brain, the cell must spill neurotransmitter across a synaptic cleft. Presumably, in order to do this, sufficient changes in the activity of calcium ion must occur within the taste cell. It is probably this change in intracellular calcium ion that is the ultimate signal arising from each transduction sequence.

II. SALTY TASTE TRANSDUCTION

Very few chemicals elicit a purely salty taste. Two of the most recognized stimuli that do impart a purely salty taste are NaCl and LiCl (Murphy, Cardello, & Brand, 1981). This stimulus specificity can be rationalized through an appreciation of the receptor mechanisms for this modality.

Saltiness (the taste of NaCl) is likely transduced by an epithelial-type ion channel located on the plasma membrane of the taste receptor cell (Avenet & Lindemann, 1988; DeSimone, Heck, & DeSimone, 1981; Heck, Mierson, & DeSimone, 1984; Schiffman, Lockhead, & Maes, 1983). In many animal models, the transduction of the signal for saltiness can be inhibited by the diuretic amiloride and some of its analogs. The existence of amiloride-inhibitable ion channels on taste cells has been inferred from psychophysical studies (McCutcheon, 1992; Schiffman et al., 1983; Tennissen & McCutcheon, 1996) and from recordings of the innervating sensory nerves (Brand, Teeter, & Silver, 1985; DeSimone & Ferrell, 1985; Hettinger & Frank, 1990; Schiffman et al., 1983; Schiffman, Suggs, Cragoe, & Erickson, 1990), and they have been directly observed using the patch–clamp technique on isolated taste cells and isolated taste buds (Avenet & Lindemann, 1988). The neural response to NaCl can also be enhanced by agents known to affect the activity of the amiloride-sensitive sodium channel, including bretylium (Schiffman, Simon, Gill, & Beeker, 1986) and novobiocin (Feigin et al., 1994). Although amiloride sensitivity of NaCl is readily demonstrable in most rodents, it appears less reliable in humans.

Several studies have failed to observe an effect of amiloride on saltiness (e.g., Desor & Finn, 1989; Ossebaard & Smith, 1995). There also appear to be marked individual differences with regard to the effectiveness of amiloride on human perception of saltiness. The lack of effect of amiloride in humans does not mean that salty taste is not transduced in humans via an epithelial sodium ion channel. Rather, the implication may be that in some individuals, the sodium channel in taste cells is less amiloride sensitive than it is in rats, or than is the comparable channel in other epithelia, such as in the kidney.

A partial clone showing homology to an amiloride-sensitive sodium channel has been cloned out of a cDNA library developed from circumvallate papillae of rat (Li, Blackshaw, & Snyder, 1994). The hybridization studies showed the presence of message in both taste and nontaste cells of the lingual epithelium. Although this may represent the clone of the salty taste receptor, the choice of circumvallate papillae as the starting point for the library was not optimal, because it is known that NaCl stimulates the glossopharyngeal nerve in a nonamiloride-sensitive manner (Formaker & Hill, 1991), suggesting that amiloride-sensitive sodium channels may not be important for salty transduction by taste cells of the circumvallate. In addition, stimulation of single circumvallate papilla by NaCl leads to sensations of primarily sourness and bitterness, with some saltiness, whereas similar stimulation of fungiform taste papillae gives rise to primarily salty sensations (Sandick & Cardello, 1983). NaCl stimulation via the chorda tympani, which innervates taste cells of the fungiform papillae, is amiloride sensitive in some animals.

The process of transduction occurs when Na^+ in the oral cavity reaches a threshold level, that is, the activity of Na^+ in the oral cavity begins to exceed that of the intercellular space of the receptor cell. In the human, this level is near 10 mM. Sodium ions then pass into the receptor cell through open epithelial ion channels, and this influx of positive charge depolarizes the cell membrane. It is likely that this depolarization then activates other voltage-sensitive ion channels within the membrane, leading ultimately to changes in ion balance across the membrane and to an increase in calcium ion activity within the cell (Figure 2A). An increase in the release of synaptic vesicular contents results, and the innervating sensory nerve changes its firing rate.

Because of the specificity of the ion channel that transduces saltiness, it has been difficult to discover or design acceptable substitutes for NaCl. In the human it appears as though the ion channel is fairly specific for Na^+ and Li^+. The chloride salt of potassium also tastes at least partly salty to some individuals, and the variation in perceived saltiness and bitterness for KCl within the population may reflect the presence of genetically determined variation (Bartoshuk, Rifkin, Marks, & Hooper, 1988). The difficulty in finding acceptable NaCl taste substitutes has led to searches for enhancers of salty taste. These agents would ideally have no taste themselves, but be able to enhance the perceived intensity of a given concentration of NaCl, presumably by extending the open time of the salty taste ion channel. Although pharmacologic agents that enhance salty taste (presumably by enhancing Na$^+$ flux across the transporting lingual epithelia) are known (Feigin et al., 1994; Lee, 1993; Schiffman et al., 1986), no agents have yet met the organoleptic and safety requirements that would make them acceptable salty taste enhancers.

Sodium salts with anions other than chloride are less salty than those of chloride. This phenomenon has been explained by noting the difference in the ability of various anions to penetrate the epithelial barrier (Ye, Heck, & DeSimone, 1991). Because chloride readily penetrates the barrier by moving through the paracellular space, it creates a more negative region at the basal portion of the taste cell, promoting influx of positive charge. This additional influx of positive charge strengthens the signal for saltiness, apparently by the further influx of Na$^+$ into the receptor cell through amiloride-sensitive Na channels located in the apical and basal regions of the taste bud. Other larger anions, less capable of penetrating the barrier, are responsible for less Na$^+$ crossing the barrier. As a result, these salts taste less intensely salty than NaCl.

III. SOUR TASTE TRANSDUCTION

Sourness is related to the presence of hydrogen ion in solution. A precise understanding of the relationship between hydrogen ion concentration, the type of anion, and the perceived intensity of the acid is still lacking. It is well appreciated, for example, that pH is not a predictor of the sour intensity for organic acids. Rather, it appears as though titratable acidity of the acid is a major predictor of sour intensity (Beatty & Cragg, 1935; Christensen, Brand, & Malamud, 1987; Norris, Noble, & Pangborn, 1984). These observations suggest that the receptor may be viewed, in the simplest terms, as a proton counter. Consequently, the search for biophysical mechanisms of sour taste has concentrated on finding those features of taste cell responses that are sensitive to the proton. At least three mechanisms have been proposed to account for hydrogen ion-induced changes in taste cell response. Each of these makes use of ion channels on taste cell plasma membranes.

The first of these involves a proton block of an outward-going K channel (Kinnamon, Dionne, & Beam, 1988; Teeter, Sugimoto, & Brand, 1989). This mechanism was described on taste cells of the mudpuppy and tiger salamander. In the mudpuppy, specific types of K channels are clustered at the apical surface of the cell, and thus the stimulation of only this small portion of the cell can effectively close enough of the K channels of this type to allow the cell to undergo a depolarization. It is not known if this mechanism is important for sour taste transduction in humans because the relevant experiments have not been performed. However, many

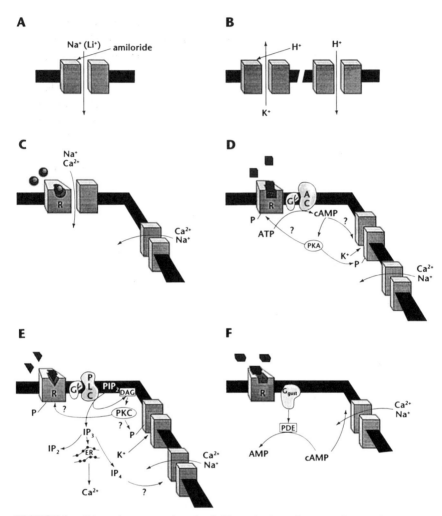

FIGURE 2   Schematic representation of probable mechanisms of taste transduction. Taste receptor elements are located at the apical processes of taste receptor cells. It is likely that the entire plasma membrane of the taste cell contains receptor elements, but, for nonpermeating stimuli, only those receptor elements above the tight junction level will be activated. In this scheme, the solid band represents the plasma membrane of the receptor cell. Elements of the transduction processes, such as receptor proteins, ion channels, and proteins of second messenger systems, are located either within this membrane or within the cytoplasm. (A) One mechanism of salty taste transduction. An epithelial-type sodium channel is located within the membrane of the receptor cell. This channel allows the passage of Na ions and Li ions from the oral environment into the cell, thereby initiating the depolarization of the receptor cell. In many species, this response is inhibitable by the diuretic, amiloride. This block of salty taste by amiloride is evidence for salty taste being mediated by an epithelial sodium channel. (B) Two mechanisms of sour taste transduction. In one scheme, protons block outward-going potassium channels, bringing about a depolarization through a buildup of positive charge in the cell. In a second scheme, protons enter the cell through an epithelial ion channel. (C) A stimulus-gated ion channel transduction scheme. In this

outward-going K channels are known to be blockable by protons, and if sufficient numbers of these are concentrated at the apical end of the taste cell, or if a sufficient number of protons can penetrate the paracellular space to interact with K channels in the basal region, it is likely that this mechanism may also function in mammals (Figure 2B).

The second mechanism recently proposed as a sour taste transduction sequence involves the proton influx through amiloride-sensitive epithelial ion channels, similar to, if not identical with, those channels that mediate salt taste. Evidence for this

---

mechanism, the stimuli (balls) bind to a recognition site on a receptor (R), which is contiguous with or closely associated with an ion channel. When the stimulus binds to the receptor site, the ion channel opens and (in the example shown here) positive charge flows into the cell. This influx of positive charge brings about a depolarization, which, if sufficient, could trigger the opening or closing of voltage-dependent ion channels in the basolateral portion of the cell. In this scheme, channels are opened and Na^+ and Ca^{2+} flow into the cell, leading to further depolarization and release of neurotransmitter. In addition to the stimulus-gated ion channel allowing influx of positive charge, this mechanism is also active where a particular stimulus interacts with a receptor that blocks outward-going potassium ions. This mechanism also leads to a depolarization. These stimulus-gated ion channel-type mechanisms are hypothesized as the transductive processes for some bitter-tasting stimuli, some sweet-tasting stimuli in selected species, the umami taste of glutamate in mammals, and the recognition of L-arginine and L-proline in the catfish taste system. (D) A second messenger system involving the stimulus-induced production of cyclic adenosine $3',5'$-monophosphate (cAMP). In this mechanism the stimulus (boxes) binds to a receptor (R), which then activates a G protein (G), which, in turn, activates the enzyme adenylyl cyclase (AC). This enzyme produces the second messenger, cAMP. This second messenger can activate protein kinase A (PKA), which phosphorylates proteins, including ion channels and possibly the receptor, activating some and inhibiting others. This mechanism usually leads to a depolarization, which, if sufficient, triggers release of neurotransmitter. The question of whether cAMP directly gates an ion channel in taste receptor cells that produce cAMP in response to a stimulus is still being evaluated. This scheme has been suggested as a mechanism of sweet taste transduction, particularly for carbohydrate sweeteners. (E) A second messenger mechanism involving the stimulus-induced production of the lipid-derived messengers, inositol 1,4,5-trisphosphate (IP_3) and diacyl glycerol (DAG). In this scheme, the stimuli (triangles) bind to a receptor (R), which activates a G protein (G). This G protein then activates the enzyme, phospholipase C (PLC), which catalyzes the conversion of the minor membrane lipid, phosphatidylinositol 4,5-bisphosphate (PIP_2), into the two second messengers, IP_3 and DAG. The messenger IP_3 may release calcium from intracellular stores, such as the endoplasmic reticulum (ER). The messenger DAG may activate protein kinase C (PKC), which may phosphorylate available sites on ion channels or receptors, activating or inhibiting them. The resulting buildup of positive charge and the increase in calcium ion activity trigger release of neurotransmitter. This scheme has been suggested as a transduction sequence for certain bitter-tasting stimuli and may be active in transducing some artificial sweeteners. (F) A second messenger mechanism involving the stimulus-induced catabolism of the second messenger cAMP. In this scheme, the stimuli (rhombi) bind to a receptor (R), which then activates the G protein G_{gust}, the taste cell-specific G protein, which, from its sequence homology to the transducins, would indicate that it functions as an activator of phosphodiesterase (PDE). The enzyme PDE catalyzes the breakdown of the second messenger cAMP to adenosine monophosphate (AMP). In order for this breakdown to lead to a depolarizing signal, cAMP may directly gate a cyclic nucleotide-inhibitable cation channel. Such a channel has been detected in taste cells of the frog. The release of inhibition due to cAMP allows these channels to open, resulting in influx of positive charge and membrane depolarization.

mechanism has accumulated primarily from studies performed with isolated taste buds of the hamster (Gilbertson, Avenet, Kinnamon, & Roper, 1992). Although the authors suggest that this may be a mechanism for sour taste transduction, it is equally likely that they are observing the passage of protons through the salty taste channel, and that this proton influx has little if any fundamental implications for sour taste transduction. It is known from the literature on amiloride-sensitive sodium channels that protons, in many cases, readily traverse these channels (Garty & Benos, 1988; Palmer, 1982).

To strengthen this hypothesis, one might look to studies that show confusions in the psychophysical distinctions between saltiness and sourness, and studies that demonstrate that one type of stimulus might enhance or inhibit the other. Although there is evidence for this confusion, it is not a large effect (Sandick & Cardello, 1983; Settle, Meehan, Williams, Doty, & Sisley, 1986). In addition, there should be an effect of amiloride on the perception of sourness. In two studies, this was not observed in humans (Schiffman et al., 1983; Tennissen & McCutcheon, 1996). In the rat and hamster, single-fiber studies do not show an amiloride effect on acid-sensitive (H) fibers (Feigin et al., 1994; Hettinger & Frank, 1990; Ninomiya & Funakoshi, 1988), but partial inhibition on sodium-sensitive (N) fiber response was observed when the tongue was being stimulated by HCl (Hettinger & Frank, 1990; Ninomiya & Funakoshi, 1988). In order to better approach this hypothesis, a systematic effort should be made to screen a large number of N fibers for their sensitivity to sour stimuli and a large number of H fibers for their sensitivity to the postulated effect of amiloride on sour stimuli. One also needs to be cognizant of species differences regarding the action of amiloride and to be aware that concentration and duration of application of the drug can influence the results of any given experiment (discussed by Hettinger & Frank, 1990). Until further physiological studies are performed, the question of the relevance of the amiloride-sensitive pathway for sour taste transduction must remain open.

A third mechanism of sour taste transduction was described from experiments in the frog (Miyamoto, Okada, & Sato, 1988; Okada, Miyamoto, & Sato, 1994). In these studies, it was observed that the cellular receptor potentials induced by acids were reduced when calcium ion concentration in the superficial fluid was reduced, or when calcium channel antagonists were present on the tongue. This response was not amiloride sensitive. The mechanism proposed is one where acid stimuli (presumably protons) activate a stimulus-gated calcium ion channel that facilitates the passage of divalent and monovalent ions into the receptor cell. The pharmacology of this mechanism is at present unique to the frog. Further work with other species is needed to determine the generality of this acid response.

Because sourness intensity is generally predictable by titratable acidity and not by pH, it is likely that the receptor mechanism for sourness involves some type of proton counting or titration. In an early review, Beidler and Gross (1971) considered this possibility and suggested that the receptor sites for sourness may be anionic moieties of the cell membrane whose state of condensation is sensitive to charge density. Intracellular pH was monitored in isolated rat taste buds undergoing stimulation with changes in external pH (Lyall, Feldman, Heck, & DeSimone, 1996). It was observed that the cells tracked the change in extracellular pH by making an analogous shift in intracellular pH. This shift to a new intracellular pH did not recover, but remained at its new level until the stimulus was removed. This pH shift was not amiloride sensitive.

IV. SWEET TASTE TRANSDUCTION

A number of chemicals with a seemingly wide array of structures can impart a sweet taste. There appear to be some basic structural determinants that can be used to predict sweetness of compounds, and by following these rules within a given chemical class, it is possible to design "supersweeteners" (see, Walters, Orthoefer, & DuBois, 1991; Nofre & Tinti, 1996). Although such rules for predicting sweetness from chemical structure would imply the existence of sweet taste receptors, there are, nevertheless, compounds that impart a sweet taste but do not fit conveniently into the accepted chemical structural framework. One example is lead acetate, another is chloroform. In addition, it is appreciated that each enantiomer of almost every monosaccharide tastes sweet. These discrepancies lead to the consideration that sweetness may be at least partly due to an interaction of some stimuli with proteins that may not be unique to the taste receptor cell, but may, nevertheless, be able to participate in a transduction sequence that ultimately leads to second messenger production or to sufficient change in the intracellular ionic environment to allow for neurotransmitter release (for discussion, see Simon, 1991; DuBois, 1997). One such possible protein could be a guanosine triphosphate (GTP)-binding protein of the taste receptor cell, and a recent intriguing paper (Naim, Seifert, Nurnberg, Grunbaum, & Schultz, 1994) lends credence to this hypothesis. Other such proteins could be ion channels.

Yet the structural correlations seen among many sweet-tasting stimuli invite the hypothesis that unique receptors do exist for sweet stimuli, although the number of these for sweet taste in humans is a matter under debate (Breslin, Kemp, & Beauchamp, 1994). Recent biochemical and biophysical studies allow mechanistic hypotheses for sweet taste transduction involving second messengers (Naim, 1993; Brand & Feigin, 1996).

The first hypothesis that was generated for sweet taste transduction using a second messenger pathway stated that sweet taste transduction occurs via a receptor/G-protein/second-messenger-coupled sequence leading to the accumulation of the second messenger, cyclic AMP. Biochemical studies have shown that sweet-tasting carbohydrates enhance the production of this second messenger in a GTP-dependent manner in tissue derived primarily from those areas of the tongue containing taste receptors (Naim, Ronen, Streim, Levinson, & Zehavi, 1991; Streim, Pace, Zehavi, Naim, & Lancet, 1989; Streim, Naim, & Lindemann, 1991). These stimuli are generally less effective at inducing cyclic AMP production in lingual epithelia devoid of taste receptors. [One interesting exception to this was the action of the stimulus, saccharin. Yet it is now known that saccharin, along with some other sweeteners, can directly stimulate the G protein (Naim et al., 1994)]. As with other signal transduction systems that produce cyclic AMP, the cyclic AMP thus produced in the taste cell likely stimulates a protein kinase A, which may phosphorylate ion channels, leading ultimately to cellular depolarization (most probably by reducing an outward current), an increase in intracellular calcium ion activity, and release of neurotransmitter (Avenet, Hofmann, & Lindemann, 1988; Tonosaki & Funakoshi, 1988) (Figure 2D). In isolated single taste bud cells that responded to sweet stimuli with a reduction in outward currents, permeable cyclic nucleotide analogs also reduced these outward currents. Additionally, sweet responses showed a cross-adaptation with nucleotide-induced responses (Cummings, Daniels, & Kinnamon, 1996). These studies with isolated taste cells support a primary role for cyclic nucleotides acting as second messengers in sweet taste transduction.

Some recent studies have suggested that the second messenger inositol 1,4,5-trisphosphate (IP_3) may also be formed during stimulation by sweet-tasting compounds (Bernhardt, Naim, Zehavi, & Lindemann, 1996). The data show that the nonsugar sweeteners, saccharin and the guanidine sweetener, SC-45647, induce the production of IP_3 when an homogenate of the epithelium from the circumvallate region of the rat is used as the tissue source. Sucrose did not induce accumulation of IP_3. Using calcium imaging of isolated taste cells, these authors also reported that the same cell could be stimulated by saccharin, SC-45647, and sucrose, yet the increases in calcium activity induced in this cell by these stimuli derived from different sources. In the case of sucrose stimulation, the intracellular increase in calcium was primarily a result of calcium ion influx from the extracellular fluid. In the case of saccharin and SC-45647, the increase in intracellular calcium was primarily a result of release of calcium from intracellular stores. These calcium imaging results are consistent with there being two different transduction sequences for sweetness in the rat. These are intriguing results, and could be strengthened by simultaneous measurements of the production of both IP_3 and cAMP in the millisecond time frame.

Recently Wong, Gannon, & Margolskee (1996) have shown that when the taste-specific G protein, gustducin, (McLaughlin, McKinnon, & Margolskee, 1992) (see Section VIII, G Proteins) is absent from mice, the animals display decreased sensitivity for sweet-tasting compounds. This observation would suggest that sweet taste is transduced by a receptor/gustducin modulated decrease in cyclic nucleotide concentration (e.g., cAMP). A decrease in cAMP is postulated since gustducin has considerable sequence homology with the transducins, G proteins of the visual system. It is known that the transducins activate a phosphodiesterase that, in turn, decreases the concentration of cGMP. Additional studies are required to address the two apparently opposing results on sweet taste and cAMP metabolism.

Other receptors for sweetness may exist that are not necessarily associated with second messenger production. For example, the existence of a stimulus-gated type of ion channel for sweet-tasting carbohydrates and artificial sweeteners can be inferred from ion transport studies on lingual epithelia and from psychophysical studies in humans (Mierson, DeSimone, Heck, & DeSimone, 1988; Schiffman et al., 1983; Simon, Labarca, & Robb, 1989). In both types of studies, the ability of the diuretic, amiloride, to suppress in the one case, ion transport, and, in the other, sweet taste, suggested the existence of a receptor/(sodium) ion channel complex for sweet taste. No other direct evidence for this type of receptor in sweet taste transduction has been published, but other stimulus-gated ion channel-type receptors in taste have been demonstrated or inferred for the recognition of L-arginine and L-proline by catfish (Brand, Teeter, Kumazawa, Huque, & Bayley, 1991; Caprio et al., 1993; Teeter et al., 1992) and L-glutamate by the mouse (Brand et al., 1991; Teeter et al., 1992).

V. BITTER TASTE TRANSDUCTION

A wide variety of compounds, both organic and inorganic, of varying structure, impart a bitter taste. The observations on this wide diversity of stimulus types are the impetus for the argument that more than one transduction mechanism exists for bitter taste. In addition, behavioral genetic studies with various strains of mice indicate that bitter taste may be mediated by several genes, because strains of mice are differentially sensitive to a number of bitter compounds. These studies are reinforced by recent biochemical and neurophysiological studies that are also pointing to the existence of a number of possible bitter taste transduction mechanisms. The possibility that secretory proteins unique to the von Ebner gland may play a role in bitter taste either by sequestering bitter stimuli or by participating directly in the taste process has also been suggested (Schmale, Holtgreve-Grez, & Christiansen, 1990; Schmale, Ahlers, Blaker, Kock, & Spielman, 1993). A symposium on bitter taste, published as a separate issue of the journal *Physiology & Behavior* (Beauchamp, 1994), summarizes many of these arguments.

Both taste cells and partial membrane preparations isolated from the circumvallate and foliate regions of mouse, rat, and bovine tongue have been used to study the transductive processes in bitter taste. Taste cells possess voltage-activated outward currents that have been identified as delayed rectifier potassium currents. In one study, about half of these cells also displayed rapidly inactivating voltage-dependent inward currents (Spielman et al., 1989). Strong suppression of the outward current was observed with application of the bitter stimulus, denatonium, at 10 μM. Whether this block was due to a direct action of denatonium on the potassium channels of the cells, or to denatonium-induced changes in intracellular levels of a second messenger(s), leading ultimately to block the outward current, was not discernible in this study. However in another study, denatonium was shown to increase intracellular calcium ion in some taste cells of the rat, even when these cells were maintained in zero extracellular calcium (Akabas, Dodd, & Al-Awqati, 1988). It was hypothesized that this increase was brought about by interaction of the bitter stimulus with a surface receptor, followed by a change in intracellular second messenger production leading to release of calcium from intracellular stores.

Several studies implicate polyphosphoinositide involvement in the receptor/second messenger transduction scheme for bitter taste. One study demonstrated the presence of the requisite enzymatic machinery in the taste bud and reported low but significant enhancement of IP_3 accumulation with stimulation by denatonium (Hwang, Verma, Bredt, & Snyder, 1990). In other studies, stimulation of a partial membrane preparation derived from circumvallate and foliate regions of mouse tongue with bitter substances, including denatonium and sucrose octaacetate, led to a GTP-dependent increase in the production of the second messenger, IP_3 (Spielman, Huque, Whitney, & Brand, 1992; Spielman, Huque, Nagai, Whitney, & Brand, 1994). The production of this second messenger was inhibited by pertussis toxin, but not by cholera toxin. Its accumulation can be observed within the millisecond time frame, with peak production occurring at 75 to 100 msec (Spielman et al., 1996).

The results from these diverse studies allow the development of a hypothetical mechanism for bitter taste transduction involving putative receptors, G proteins, and the generation of polyphosphoinositol-derived second messengers (Figure 2E). This model assumes the presence of receptors of as yet unidentified character. These receptors are linked with a G protein, possibly a $G_{\alpha i1}$, a $G_{\alpha i2}$, or a G_q. When stimuli bind to the receptor, the α subunit of the G protein activates a phospholipase C, which then catabolizes membrane-associated phosphatidylinositol-4,5-bisphosphate (PIP_2) into two second messengers, IP_3 and diacyl glycerol (DAG). The IP_3 could release calcium ion from intracellular stores (its traditional role) or could be further phosphorylated to IP_4, which in some cells is a stimulator of plasma membrane-associated calcium channels, allowing influx of calcium from extracellular spaces. DAG, as in many other cells, may activate the enzyme, protein kinase C, which can phosphorylate available sites. Some of these sites may be on ion channels, which may then be activated or inhibited; others may be on the receptors themselves, causing down-regulation of these. Cellular depolarization may result from activation of ion channels, allowing influx of positive charge, and/or from inhibition of outward-going potassium channels, leading to build-up of positive charge within the cell. The depolarizing event will activate voltage-sensitive calcium channels and the IP_3-induced release of intracellular calcium may also activate calcium-sensitive calcium channels, and can also activate phospholipase C (PLC) and induce further release of sequestered calcium. These events will lead to sufficient increase in intracellular calcium ion activity, triggering neurotransmitter release. The response is terminated by metabolism of the second messengers, by dephosphorylation events, and by removal and/or sequestering of the surfeit of intracellular ions.

While the receptor/second messenger transduction pathway has credibility, it is also true that a number of well-known potassium channel blockers (e.g., tetraethylammonium, quinine) are also bitter tasting, suggesting that the ability of these compounds to block directly different types of potassium channels may be an important mechanism of bitter taste transduction. The closure of potassium channels appears to be a common feature for taste cells being stimulated by a variety of taste-active compounds (Kinnamon, 1992; Cummings et al., 1996; Spielman et al., 1989). Cloning studies have resulted in the identification of several potassium channels in a library made from circumvallate papillae of rats (Hwang, Glatt, Bredt, Yellen, & Snyder, 1992). This identification of channels allows their expression in heterologous systems, leading to a better understanding of their activity in taste transduction.

A third transduction mechanism has been proposed for bitter taste involving the taste cell-specific G protein called gustducin (McLaughlin, McKinnon, & Margolskee, 1992). The α subunit of this G protein shares considerable sequence homology with the well-known G proteins of the visual system, the transducins. It is assumed, therefore, that the activity of gustducin is similar to that of the transducins. It is known that the transducins stimulate the activity of the phosphodiesterase enzyme that metabolizes cyclic nucleotides to their monophosphate form (e.g., cGMP to GMP). It is assumed that gustducin will be found to have similar specificity. If the gustducin cascade is involved in bitter taste, then the transduction sequence will likely be found to be as follows (Figure 2F): The stimulus interacts with a bitter taste receptor that causes the gustducin G protein to break apart, allowing the α subunit of gustducin to stimulate phosphodiesterase. This stimulation reduces the concentration of cyclic nucleotides. In order for this reduction in cyclic nucleotides to lead to an excitatory event, the cyclic nucleotide must be gating an inhibitory channel, so that when its concentration decreases, a membrane-depolarizing event can occur. Such a cyclic nucleotide-suppressible channel has been observed in frog taste cells (Kolesnikov & Margolskee, 1995). This release of inhibition on the channel would allow calcium ion to enter the cell and neurotransmitter to be released.

Perhaps the gustducin sequence and the inositol phosphate sequence are not mutually exclusive, but rather related. For example, assume the receptor recognizes a stimulus such as denatonium. This occupied receptor may then interact with the taste-specific G protein, gustducin. The α subunit breaks away and activates phosphodiesterase. As in some other signaling systems, the other portion, the β/γ subunit, may simultaneously activate PLC. Phospholipase C then produces IP_3 and DAG. IP_3 releases calcium from intracellular stores. This calcium then further activates phosphodiesterase (PDE), a calcium-dependent enzyme. The PDE then metabolizes cAMP, leading to opening of ion channels. DAG activates protein kinase C, which phosphorylates available sites, including those on enzymes, receptors, and ion channels that will at first augment the signal and later help to dampen it. This hypothesis, as yet untested, explains some recent data that show denatonium inducing both an IP_3 increase and a PDE activation (Ruiz-Avila et al., 1995; Spielman et al., 1996). Also, in recent single-cell studies, denatonium was found to bring about intracellular calcium increases through both release of calcium from internal stores (an IP_3 event) and influx of calcium from extracellular space (perhaps a release of the cyclic nucleotide-suppressible channel) (Ogura, Mackay-Sim, & Kinnamon, 1996).

VI. UMAMI TASTE TRANSDUCTION

The debate over whether umami taste is a fifth basic taste is likely to continue for some time. This discussion may be clarified by studies that seek to find the transduction mechanisms for umami taste. The prototypical stimulus for this taste is the monosodium salt of L-glutamic acid (monosodium glutamate; MSG). Studies searching for the transduction mechanism of umami taste have centered on hypotheses about the receptors for L-glutamate.

There are a variety of neurotransmitter receptors for glutamate in the central nervous system (Hollmann & Heinemann, 1994). They fall into two category types: stimulus-gated ion channels and the metabotropic receptors, that is, those that stimulate the production or catabolism of second messengers. Studies of the taste receptor for glutamate should, like their counterparts in the CNS, strive to make use of the unique pharmacology of these systems. In taste, for example, certain 5′-ribonucleotides can enhance the intensity of the taste of MSG in a true taste synergy (Rifkin & Bartoshuk, 1980; Yamaguchi, 1967; Yamaguchi, Yoshikawa, Ikeda, & Ninomiya, 1971). The successful characterization of the umami taste receptor must, therefore, include a functional explanation for this synergy.

The first biochemical studies to characterize the receptor for glutamate in taste tissue reported that L-glutamate bound in a tissue-specific manner to a partial membrane preparation from bovine circumvallate tissue, and that this binding could be enhanced by the presence of certain 5′-ribonucleotides (Torii & Cagan, 1980). The nature of the transductive step remained unexplored until recently.

To date, two hypotheses have been put forth that describe the taste receptor for glutamate. One states that the receptor is a stimulus-gated ion channel receptor. This hypothesis was initially supported by studies showing that a glutamate-stimulated ion channel could be reconstituted into a lipid bilayer from a partial membrane preparation of mouse circumvallate tissue. In otherwise silent bilayers, addition of MSG led to an increase in conductance of the bilayer (Brand et al., 1991; Teeter et al., 1992). This conductance increase was graded with MSG concentration and could be enhanced by the addition of 5′-guanosine monophosphate (GMP), a known enhancer of the MSG response in the mouse. These findings suggested that the taste receptor for MSG in the mouse may be of the stimulus-gated ion channel type, perhaps similar to the N-methyl-D-aspartate (NMDA) receptor (Figure 2C).

The second hypothesis states that the umami receptor is a metabotropic glutamate receptor (Chaudhari et al., 1996). This hypothesis received support when it was reported that a cDNA library, constructed from rat circumvallate tissue, included sequences similar to known CNS glutamate receptors. Several clones were found. *In situ* hybridization studies showed that one of these, a low-abundance clone of the metabotropic-like receptor, mGluR4, could be localized specifically to the taste buds of rat circumvallate. These data, therefore, suggest that the receptor for umami may be a metabotropic-like glutamate receptor.

If the umami receptor is a metabotropic receptor, then it should be possible to observe the production or catabolism of second messenger while tissue from the circumvallate is being stimulated with MSG. One attempt to measure second messenger production reported no accumulation of either IP_3 or cAMP above background during stimulation with MSG (Brand et al., 1991). However, if the receptor in an mGluR4, then stimulation by glutamate and agonists for mGluR4 should lead to decreases in cAMP.

It has been reported that isolated taste cells from mouse vallate respond to L-glutamate with both increases and decreases in intracellular calcium (Hayashi, Zviman, Brand, Teeter, & Restrepo, 1996). In contrast, the analog N-methyl-D-aspartate elicited only increases in intracellular calcium accompanied by a depolarization, whereas the analog L-2-amino-4-phosphonobutyrate (L-AP4) elicited primarily decreases in intracellular calcium. NMDA is an agonist for a stimulus-gated ion channel receptor for glutamate, whereas L-AP4 is an agonist for a metabotropic glutamate receptor. These data suggest, therefore, that there may be at least two types of glutamate receptors in taste cells: an excitatory receptor of the stimulus-gated ion channel type and a receptor of the metabotropic type. Because taste cells signal the presence of stimuli with excitatory responses, it is likely that the receptor for umami is a stimulus-gated ion channel. However, it is also possible that the metabotropic receptor functions in taste transduction, because inhibitory responses could be important in taste processing. It is also possible that the umami receptor is one of a subset of metabotropic receptors for glutamate but that its abundance is too low to allow its excitatory responses to be observed using the conditions of this study. Clearly the receptor and transduction processes for umami are just beginning to be explored.

VII. THE QUESTION OF RECEPTORS

Protein receptors unique to the taste system are just beginning to be isolated and cloned. Although the probability that receptors exist for certain modalities is strong, no candidates have been directly and functionally linked to taste transduction. For certain modalities, including sweet, bitter, and umami taste in mammals and amino acid taste in fish, there is strong circumstantial evidence for unique receptors. This is particularly true in the well-studied animal model, the channel catfish (Caprio et al., 1993). This animal is sensitive to a number of amino acids, and both electrophysiological and biochemical/biophysical studies suggest that at least three separate receptor systems are involved in this recognition. A number of studies point to partial purification of a taste receptor for L-arginine in this animal (Kalinoski, Spielman, Teeter, Andreini, & Brand, 1994; Grosvenor et al., 1996). This taste receptor for L-arginine is a stimulus-gated ion channel. The channel catfish also possesses a second stimulus-gated ion channel-type taste receptor for L-proline. In contrast, the receptor for alanine and other neutral amino acids appears to be linked via a G protein to second messenger formation (Brand & Bruch, 1992; Caprio et al., 1993). Recent studies using probes against receptor activity suggest that the majority of receptor cells express one receptor/protein type (Finger et al., 1996). Studies such as these with this favorable model system suggest the existence of unique receptor proteins for taste and imply similar receptor/transductive systems in mammalian models.

Because the salt taste modality apparently utilizes an epithelial sodium ion channel, it would appear to be relatively straightforward to apply the cloning techniques used in identifying other epithelial sodium channels (e.g., Canessa, Horisberger, & Rossier, 1993) to lingual taste channels. Such an effort has been reported. The clone so identified localizes both to cells of the taste buds and to other lingual epithelial cells (Li et al., 1994). The potential problems with this study were discussed in Section II. It would be prudent to repeat these studies with a cDNA library derived from fungiform taste tissue, because these cells, innervated as they are by the chorda tympani, are known to display sensitivity to a number of epithelial-type sodium channel agonists and antagonists, including amiloride (and some of its derivatives), bretylium, and novobiocin.

Direct purification of taste receptors from taste tissue of mammals has not met with success. Based on the success at identifying the clones of olfactory receptors (Buck & Axel, 1991; Lancet & Ben-Arie, 1993), several groups have undertaken similar searches in taste and lingual cDNA libraries, confining their investigation to clones that might display the classic seven transmembrane domain-type sequences. Several groups have reported results from these approaches. The reported sequences show homology with the known olfactory receptor sequences (Abe, Kusakabe, Tanemura, Emori, & Arai, 1993; Matsuoka, Mori, Aoki, Sato, & Kurihara, 1993), and in situ hybridization studies show that their expression is not unique to taste receptor cells. Another study reported the identification of a neuropeptide-like receptor expressed not only in taste tissue but also in some other sensory tissues (Tal, Ammar, Karpuj, Krizhanovsky, Naim, & Thompson, 1995). Its function in taste is not known.

It is likely that taste cell-specific receptors exist. Given the low abundance of taste cells and the probable low level of expression of receptors in these cells, their identification may prove to be a daunting, but nevertheless rewarding, task.

VIII. G PROTEINS

Receptors are often coupled with a special class of heterotrimeric regulatory proteins called GTP-binding proteins, or G proteins (Hepler & Gilman, 1992; Hille, 1992). These proteins are responsible for transferring the energy derived from the interaction of the receptor with its agonist into intracellular signals. In many cases, these signals are characterized by changes in the levels of second messengers. When a receptor, occupied by its agonist, interacts with a G protein, the G protein dissociates. One part of the G protein, the α subunit, is generally responsible for activating the enzymes that alter the levels of intracellular messengers. The remaining complex, the β/γ subunit, also has physiological roles in the cell, which are only now being recognized (Clapham & Neer, 1993).

Normally, G proteins are thought of as intermediaries between the receptor event and the second messenger event. Yet there is evidence that extrinsic agents can interact directly with G proteins and thereby affect cellular processes by bypassing a receptor step. This type of direct activation of G proteins has been proposed as the means by which bradykinin and some amphiphilic peptides achieve their biological activity (e.g., Mousli, Bueb, Bronner, Rouot, & Landry, 1990). Perhaps the bitter-tasting peptides and some other taste stimuli act in a similar fashion in the taste system (Naim et al., 1994; Spielman et al., 1992). In a recent report, it was directly demonstrated that some taste stimuli can activate G_i / G_o proteins and transducin (Naim et al., 1994).

The number of known members of the G protein family is growing. Within this large family there are several subfamilies that were traditionally classified by their activity toward particular second messenger-generating enzymes. One of these subfamilies is that having sequence homology and probable functional similarities to the transducins, the G proteins first identified in the visual system. Interestingly, a clone of an α subunit of a G protein unique to the taste system has been identified as having deduced sequence similarities to the α-transducins. The G protein to which this unique α subunit in the taste system belongs has been named gustducin (McLaughlin et al., 1992). Assuming that the activity of α-gustducin will be found to be the same as that of the α-transducins, then this taste-specific G protein should be responsible for activating a phosphodiesterase, which should decrease the level of cyclic nucleotides in the taste cell (Margolskee, 1993; Ruiz-Avila et al., 1995) (Figure 2F).

There is as yet no evidence that any of the mechanisms investigated for taste might depend on a decrease in levels of cyclic nucleotides. One hypothesis held that α-gustducin is involved in transduction of some bitter-tasting stimuli (Margolskee, 1993). Yet it is interesting that α-gustducin can be localized to a subset of cells in every taste bud from all three taste-bud-containing areas of the tongue (McLaughlin et al., 1992; Tabata et al., 1995). These distribution patterns would argue for gustducin playing a larger role in taste transduction. Recent results from studies on mice possessing a deletion in the α-gustducin gene (so-called knock-out mice) showed that these animals were impaired in their ability to taste both bitter and sweet stimuli (Wong, Gannon, & Margolskee, 1996). Responses to salty and sour were not affected. This implies that gustducin plays a major role in at least two modalities.

IX. BIOPHYSICS AND BEHAVIOR: IS THERE A LINK?

The peripheral transductive processes of taste are remote from the central processing that ultimately determines behavior. Between these two extremes are many intervening integrative and screening steps that may alter, or modify, the messages inherent in the peripheral recognition process. It is appropriate to question whether there are implications for human behavior in our increasing understanding of these peripheral processes.

It is instructive to first address briefly the question using a simpler organism, namely, the catfish. This animal possesses very sensitive taste receptors, tuned primarily to certain amino acid stimuli. Its taste system has been characterized from many levels, including the behavioral, anatomical, neurophysiological, and biophysical. One study integrated knowledge from these diverse areas to determine if, for example, a detailed understanding of the biophysics of the receptors could be used to predict the neurophysiology and/or behavior of the animal (Caprio et al., 1993). In fact, some striking observations can be made. This animal possesses taste receptors for amino acids. Three of these receptors (for L-alanine and other neutral amino acids, for L-arginine, and for L-proline) have been well characterized. It appears that this triplet of receptors maintains separate "identities" through the periphery and, perhaps, even up to the behavioral level. For example L-arginine is able to release behaviors that are not stimulated by L-alanine. Likewise, as predicted from the biophysics, the D isomers of these stimuli are partial antagonists of these behaviors. In addition, the behavioral responses to L-proline are elicited only at high concentrations of this amino acid and the dose–response for the behavior due to L-proline is quite steep. This pattern has a direct correlate in transductive biophysics, which shows that the receptor for L-proline responds only to high concentrations of this amino acid. It is interesting that all three of these receptors initiate appetitive behaviors. (They are, in essence, the catfish equivalent of "sweet" receptors.)

A somewhat similar argument could be put forth for mammalian taste receptors, although the intervening neurophysiology is not as direct as it is in the catfish. In mammals, there appear to be somewhat separate receptor systems for the traditional taste modalities. Increased knowledge of these at the level of the transductive process will facilitate our understanding of the behavioral consequences of taste. For example, there is still a debate as to the number of receptors for sweetness and for bitterness. Biophysical and biochemical studies could certainly directly address questions such as these. But would the resolution of these questions be of any use to behavioral studies?

It is likely that these answers will be of use, given some of the anomalies that are seen, for example, in the way in which certain compounds are able to modulate certain modalities selectively. For example, the well-known sweet taste inhibitor, gymnemic acid, is able to block the sensation of sweetness in humans and some other Old World Primates. This inhibition is observed for all sweeteners, both artificial and natural, including the sweet carbohydrates. Yet gymnema is not effective in other mammals, even those that find carbohydrates acceptable or in those that can taste some artificial sweeteners. The implication from these observations is that there must be some fundamental difference between the sweet taste receptors of the Old World primates and those of other mammals. Likewise the debate over the number of sweet receptors would be better focused with information from the periphery. Similar arguments hold for bitter taste receptors. Certain agents, such as NaCl, are able to inhibit selected bitter compounds (Yokomukai, Breslin, Cowart, & Beauchamp, 1994; Breslin & Beauchamp, 1995). There must be a peripheral correlate to this psychophysical observation, and a more detailed knowledge of the peripheral biophysics (at the receptor, cellular, and taste bud levels) of bitter taste will undoubtedly allow a more sophisticated approach to these behavioral questions.

Ultimately there will be a blending of biophysics with behavior, and information from one approach will directly impact the other. Such a merging is now taking place in studies on certain sensitive and well-defined systems, such as those found in the catfish, where knowledge of the peripheral biophysics is being used to interpret directly and, in some cases, guide behavioral studies. It is important to attempt to attain this same integration for mammals, including humans.

Acknowledgments

The author is supported in part by NIH Grant DC-00356, and by a grant from the Department of Veterans Affairs.

References

Abe, K., Kusakabe, Y., Tanemura, K., Emori, Y., & Arai, S. (1993). Primary structure and cell-type specific expression of a gustatory G-protein-coupled receptor related to olfactory receptors. *Journal of Biological Chemistry 268*, 12033–12039.

Akabas, M. H. (1990). Mechanisms of chemosensory transduction in taste cells. *International Review of Neurobiology, 32*, 241–279.

Akabas, M. H., Dodd, J., & Al-Awqati, Q. (1988). A bitter substance induces a rise in intracellular calcium in a subpopulation of rat taste cells. *Science, 242*, 1047–1050.

Avenet, P., Hofmann, F., & Lindemann, B. (1988). Transduction in taste receptor cells requires cAMP-dependent protein kinase. *Nature, 331*, 351–354.

Avenet, P., & Kinnamon, S. C. (1991). Cellular basis of taste reception. *Current Opinion in Neurobiology, 1*, 198–203.

Avenet, P., & Lindemann, B. (1988). Amiloride-blockable sodium currents in isolated taste receptor cells. *Journal of Membrane Biology, 105*, 245–255.

Bartoshuk, L. M., Rifkin, B., Marks, L. E., & Hooper, J. E. (1988). Bitterness of KCl and benzoate: Related to PTC/PROP. *Chemical Senses, 13*, 517–528.

Beatty, R. M., & Cragg, L. H. (1935). The sourness of acids. *Journal of the American Chemical Society, 57*, 2347–2351.

Beauchamp, G. K. (Ed.) (1994). International symposium on bitter taste. *Physiology and Behavior 56(6)*, 1121–1266.

Beidler, L. M., & Gross, G. W. (1971). The nature of taste receptor sites. In W. D. Naff (Ed.), *Contributions to sensory physiology* (pp. 97–127). New York: Academic Press.

Bernhardt, S. J., Naim, M., Zehavi, U., & Lindemann, B. (1996). Changes in IP_3 and cytosolic Ca^{2+} in response to sugars and non-sugar sweeteners in transduction of sweet taste in the rat. *Journal of Physiology, 490.2*, 325–336.

Brand, J. G., & Bruch, R. C. (1992). Molecular mechanisms of chemosensory transduction: Gustation and olfaction. In T. J. Hara (Ed.), *Fish chemoreception* (pp. 126–149). London: Chapman & Hall.

Brand, J. G., & Bryant, B. P. (1994). Receptor mechanisms for flavour stimuli. *Food Quality and Preference, 5*, 31–40.

Brand, J. G., & Feigin, A. M. (1996). Biochemistry of sweet taste transduction. *Food Chemistry 56,* 199–207.

Brand, J. G., Teeter, J. H., Kumazawa, T., Huque, T., & Bayley, D. L. (1991). Transduction mechanisms for the taste of amino acids. *Physiology & Behavior, 49,* 899–904.

Brand, J. G., Teeter, J. H., & Silver, W. (1985). Inhibition by amiloride of chorda tympani responses evoked by monovalent salts. *Brain Research, 334,* 207–214.

Breslin, P. A. S., & Beauchamp, G. K. (1995). Suppression of bitterness by sodium: Variation among bitter taste stimuli. *Chemical Senses, 20,* 609–623.

Breslin, P. A. S., Kemp, S., & Beauchamp, G. K. (1994). Single sweetness signal. *Nature, 369,* 447–448.

Buck, L., & Axel, R. (1991). A novel multigene family may encode odorant receptors: A molecular basis for odor recognition. *Cell, 65,* 175–187.

Canessa, C. M., Horisberger, J. D., & Rossier, B. C. (1993). Epithelial sodium channel related to proteins involved in neurodegeneration. *Nature, 361,* 467–470.

Caprio, J., Brand, J. G., Teeter, J. H., Valentincic, T., Kalinoski, D. L., Kohbara, J., Kumazawa, T., & Wegert, S. (1993). The taste system of the channel catfish: From biophysics to behavior. *Trends in Neurosciences, 16,* 192–197.

Chaudhari, N., Yang, H., Lamp, C., Delay, E., Cartford, C., Than, T., & Roper, S. (1996). The taste of monosodium glutamate: Membrane receptors in taste buds. *Journal of Neuroscience 16,* 3817–3826.

Christensen, C. M., Brand, J. G., & Malamud, D. (1987). Salivary changes in solution pH: A source of individual differences in sour taste perception. *Physiology & Behavior, 40,* 221–227.

Clapham, D. E., & Neer, E. J. (1993). Bifurcating pathways for transmembrane signalling: New roles for G protein beta/gamma subunits. *Nature, 365,* 403–406.

Corey, D. P., & Roper, S. D. (Eds.) (1992). *Sensory transduction.* New York: Rockefeller Univ. Press.

Cummings, T. A., Daniels, C., & Kinnamon, S. C. (1996). Sweet taste transduction in hamster— Sweeteners and cyclic-nucleotides depolarize taste cells by reducing a K⁺ current. *Journal of Neurophysiology, 75,* 1256–1263.

DeSimone, J. A., & Ferrell, F. (1985). Analysis of amiloride inhibition of chorda tympani taste response of rat to NaCl. *American Journal of Physiology, 249,* R52–R61.

DeSimone, J. A., Heck, G. L., & DeSimone, S. K. (1981). Active ion transport in dog tongue: A possible role in taste. *Science, 214,* 1039–1041.

Desor, J. A., & Finn, J. (1989). Effects of amiloride on salt taste in humans. *Chemical Senses, 14,* 793–803.

DuBois, G. E. (1997). New insights on the coding of the sweet taste message in chemical structure. In G. Salvadori (Ed.) *Olfaction and taste—A century for the senses* (pp. 32–95) Carol Stream, IL: Allured Publishing.

Feigin, A. M., Ninomiya, Y., Bezrukov, S. M., Bryant, B. P., Moore, P. A., Komai, M., Wachowiak, M., Teeter, J. H., Vodyanoy, I., & Brand, J. G. (1994). Enhancement of gustatory nerve fibers to NaCl and formation of ion channels by a commercial preparation of novobiocin. *American Journal of Physiology, 266,* C1165–C1172.

Finger, T. E., Bryant, B. P., Kalinoski, D. L., Teeter, J. H., Böttger, B., Grosvenor, W., Cagan, R. H., & Brand, J. G. (1996). Differential localization of putative amino acid receptors in taste buds of the channel catfish, *Ictalurus punctatus. Journal of Comparative Neurology 373,* 129–138.

Formaker, B. K., & Hill, D. L. (1991). Lack of amiloride sensitivity in SHR and WKY glossopharyngeal taste responses to NaCl. *Physiology & Behavior, 50,* 765–769.

Garty, H., & Benos, D. J. (1988). Characteristics and regulatory mechanisms of the amiloride-blockable Na channel. *Physiological Reviews, 68,* 309–373.

Gilbertson, T. A. (1993). The physiology of vertebrate taste reception. *Current Opinion in Neurobiology, 3,* 532–539.

Gilbertson, T. A., Avenet, P., Kinnamon, S. C., & Roper, S. D. (1992). Proton currents through amiloride-sensitive Na channels in hamster taste cells: Role in acid transduction. *Journal of General Physiology, 100,* 803–824.

Grosvenor, W., Spielman, A. I., Feigin, A. M., Kalinoski, D. L., Finger, T. E., Wood, M., DellaCorte, C., Andreini, I., Teeter, J. H., & Brand, J. G. (1996). Purification of an arginine taste receptor from the channel catfish. [Abstract]. *Chemical Sense 21,* 610.

Hayashi, Y., Zviman, M. M., Brand, J. G., Teeter, J. H., & Restrepo, D. (1996). Measurement of membrane potential and $[Ca^{2+}]_i$ in cell ensembles: Applications to the study of glutamate taste in mouse. *Biophysical Journal 71,* 1057–1070.

Heck, G. L., Mierson, S., & DeSimone, J. A. (1984). Salt taste transduction occurs through an amiloride-sensitive sodium transport pathway. *Science, 233,* 403–405.

Helper, J. R., & Gilman, A. G. (1992). G proteins. *Trends in Biological Sciences, 17,* 383–387.

Hettinger, T. P., & Frank, M. E. (1990). Specificity of amiloride inhibition of hamster taste responses. *Brain Research, 513,* 24–34.

Hille, B. (1992). G protein-coupled mechanisms and nervous signaling. *Neuron, 9,* 187–195.

Hollmann, M., & Heinemann, S. (1994). Cloned glutamate receptors. *Annual Review of Neuroscience, 17,* 31–108.

Hwang, P. M., Glatt, C. E., Bredt, D. S., Yellen, G., & Snyder, S. H. (1992). A novel K^+ channel with unique localization in mammalian brain: Molecular cloning and characterization. *Neuron, 8,* 473–481.

Hwang, P. M., Verma, A., Bredt, D. S., & Snyder, S. H. (1990). Localization of phosphatidyl-inositol signaling components in rat taste cells: Role in bitter taste transduction. *Proceedings of the National Academy of Sciences, USA, 87,* 7395–7399.

Kalinoski, D. L., Spielman, A. I., Teeter, J. H., Andreini, I., & Brand, J. G. (1994). Strategies for isolation of taste receptor proteins. In K. Kurihara, N. Suzuki, & H. Ogawa (Eds.) *Olfaction & taste XI* (pp. 73–76). New York: Springer-Verlag.

Kinnamon, J. C. (1987). Organization and innervation of taste buds. In T. E. Finger & W. L. Silver (Eds.) *Neurobiology of taste and smell* (pp. 277–297). New York: Wiley.

Kinnamon, S. C. (1992). Role of K channels in taste transduction. In D. P. Corey & S. D. Roper (Eds.) *Sensory transduction* (pp. 261–270). New York: Rockefeller Univ. Press.

Kinnamon, S. C., & Cummings, T. A. (1992). Chemosensory transduction mechanisms in taste. *Annual Review of Physiology, 54,* 715–731.

Kinnamon, S. C., & Margolskee, R. F. (1996). Mechanisms of taste transduction. *Current Opinion in Neurobiology 6,* 506–513.

Kinnamon, S. C., Dionne, V. E., & Beam, K. G. (1988). Apical localization of K channels in taste cells provides the basis for sour taste transduction. *Proceedings of the National Academy of Sciences, USA, 85,* 7023–7027.

Kolesnikov, S. S., & Margolskee, R. F. (1995). A cyclic-nucleotide-suppressible conductance activated by transducin in taste cells. *Nature, 376,* 85–88.

Kurihara, K. (1990). Molecular mechanisms of reception and transduction in olfaction and taste. *Japanese Journal of Physiology, 40,* 305–324.

Lancet, D., & Ben-Arie, N. (1993). Olfactory receptors. *Current Biology, 3,* 668–674.

Lee, T. D. (1993). Seasoned food product with a salt enhancer. U.S. Patent #5,176,934.

Li, X.-J., Blackshaw, S., & Snyder, S. H. (1994). Expression and localization of amiloride-sensitive sodium channel indicate a role for non-taste cells in taste perception. *Proceedings of the National Academy of Sciences, USA, 91,* 1814–1818.

Lindemann, B. (1996). Taste reception. *Physiological Reviews 76,* 719–766.

Lyall, V., Feldman, G. M., Heck, G. L., & DeSimone, J. A. (1996). Measurement of intracellular pH (pH_i) in isolated rat circumvallate papillae taste receptor cells. [Abstract 183]. Association for Chemoreception Sciences XVIII Meeting, Sarasota, FL.

Margolskee, R. F. (1993). The biochemistry and molecular biology of taste transduction. *Current Opinion in Neurobiology, 3,* 526–531.

Matsuoka, I., Mori, T., Aoki, J., Sato, T., & Kurihara, K. (1993). Identification of novel members of G protein coupled receptor superfamily expressed in bovine taste tissue. *Biochemical and Biophysical Research Communications, 194,* 504–511.

McCutcheon, N. B. (1992). Human psychophysical studies of saltiness suppression by amiloride. *Physiology and Behavior 51*, 1069–1074.

McLaughlin, S. K., McKinnon, P. J., & Margolskee, R. F. (1992). Gustducin is a taste-cell-specific G protein subunit closely related to the α-transducins. *Nature, 357*, 563–569.

Mierson, S., DeSimone, S. K., Heck, G. L., & DeSimone, J. A. (1988). Sugar-activated ion transport in canine lingual epithelium. *Journal of General Physiology, 92*, 87–111.

Miyamoto, T., Okada, Y., & Sato, T. (1988). Ionic basis of receptor potential of frog taste cells induced by acid stimuli. *Journal of Physiology, 405*, 699–711.

Mousli, M., Bueb, J. L., Bronner, C., Rouot, B., & Landry, Y. (1990). G protein activation: A receptor independent mode of action for cationic amphiphilic neuropeptides and venom peptides. *Trends in Neurosciences, 11*, 358–362.

Murphy, C., Cardello, A. V., & Brand, J. G. (1981). Tastes of fifteen halide salts following water and NaCl: Anion and cation effects. *Physiology & Behavior, 26*, 1083–1095.

Naim, M. (1993). Cellular transduction of sugar-induced sweet taste. In G. Charalambous (Ed.) *Food flavors, ingredients and composition* (pp. 647–656). Amsterdam: Elsevier.

Naim, M., Ronen, T., Streim, B. J., Levinson, M., & Zehavi, U. (1991). Adenylate cyclase responses to sucrose stimulation in membranes of pig circumvallate taste papillae. *Comparative Biochemistry and Physiology, 100B*, 455–458.

Naim, M., Seifert, R., Nurnberg, B., Grunbaum, L., & Schultz, G. (1994). Some taste substances are direct activators of G proteins. *Biochemical Journal 297*, 451–454.

Ninomiya, Y., & Funakoshi, M. (1988). Amiloride inhibition of response of rat single chorda tympani fibers to chemical and electrical tongue stimulations. *Brain Research 451*, 319–325.

Nofre, C., & Tinti, J.-M. (1996). Sweetness reception in man: The multipoint attachment theory. *Food Chemistry 56*, 263–274.

Norris, M. B., Noble, A. C., & Pangborn, R. M. (1984). Human saliva and taste responses to acids varying in anions, titratable acidity, and pH. *Physiology & Behavior, 32*, 237–244.

Ogura, T., Mackay-Sim, A., & Kinnamon, S. C. (1996). Patch-clamp and optical studies of bitter transduction in isolated taste cells [abstract]. *Chemical Senses 21*, 278.

Okada, Y., Miyamoto, T., & Sato, T. (1994). Activation of a cation conductance by acetic acid in taste cells isolated from the bullfrog. *Journal of Experimental Biology, 187*, 19–32.

Ossebaard, C. A., & Smith, D. V. (1995). Effects of amiloride on the taste of NaCl, Na-gluconate, and KCl in humans: Implications for Na⁺ receptor mechanisms. *Chemical Senses, 20*, 37–46.

Palmer, L. G. (1982). Ion selectivity of the apical membrane Na channel in the toad urinary bladder. *Journal of Membrane Biology, 67*, 91–98.

Reutter, K. (1978). Taste organ in the bullhead (*Teleostei*). *Advances in Anatomy, Embryology and Cell Biology, 55*, 1–98.

Rifkin, B., & Bartoshuk, L. M. (1980). Taste synergism between monosodium glutamate and disodium 5′-guanylate. *Physiology & Behavior, 24*, 1169–1172.

Roper, S. D. (1989). The cell biology of vertebrate taste receptors. *Annual Review of Neuroscience, 12*, 329–353.

Roper, S. D. (1992). The microphysiology of peripheral taste organs. *The Journal of Neuroscience, 12*, 1127–1134.

Ruiz-Avila, L., McLaughlin, S. K., Wildman, D., McKinnon, P. J., Robichon, A., Spickofsky, N., & Margolskee, R. F. (1995). Coupling of bitter receptor to phosphodiesterase through transducin in taste receptor cells. *Nature, 376*, 80–85.

Sandick, B., & Cardello, A. V. (1983). Tastes of salts and acids on circumvallate papillae and anterior tongue. *Chemical Senses, 8*, 59–69.

Schiffman, S. S., Lockhead, E., & Maes, F. W. (1983). Amiloride reduces the taste intensity of Na and Li salts and sweeteners. *Proceedings of the National Academy of Sciences, USA, 80*, 6136–6140.

Schiffman, S. S., Simon, S. A., Gill, J. M., & Beeker, T. G. (1986). Bretylium tosylate enhances salt taste. *Physiology & Behavior, 36*, 1129–1137.

Schiffman, S. S., Suggs, M. S., Cragoe, E. J., Jr., & Erickson, R. P. (1990). Inhibition of taste responses to Na salts by epithelial Na channel blockers in gerbil. *Physiology & Behavior, 47,* 455–459.

Schmale, H., Ahlers, C., Blaker, M., Kock, K., & Spielman, A. I. (1993). Perireceptor events in taste. *Ciba Foundation Symposium, 179,* 167–185.

Schmale, H., Holtgreve-Grez, H., & Christiansen, H. (1990). Possible role for salivary gland protein in taste reception indicated by homology to lipophilic-ligand carrier proteins. *Nature, 343,* 366–369.

Settle, R. G., Meehan, K., Williams, G. R., Doty, R. L., & Sisley, A. C. (1986). Chemosensory properties of sour tastants. *Physiology & Behavior, 36,* 619–623.

Simon, S. A. (1991). Mechanisms of sweet taste transduction. In D. E. Walters, F. T. Orthoefer, & G. E. DuBois (Eds.) *Sweeteners. Discovery, molecular design, and chemoreception* (pp. 237–250). Washington, DC: American Chemical Society.

Simon, S. A., Labarca, P., & Robb, R. (1989). Activation by saccharides of a cation-selective pathway on canine lingual epithelium. *American Journal of Physiology, 256,* R394–R402.

Spielman, A. I., Huque, T., Nagai, H., Whitney, G., & Brand, J. G. (1994). Generation of inositol phosphates in bitter taste transduction. *Physiology & Behavior, 56,* 1149–1155.

Spielman, A. I., Huque, T., Whitney, G., & Brand, J. G. (1992). The diversity of bitter taste signal transduction mechanisms. In D. P. Corey & S. D. Roper (Eds.) *Sensory transduction* (pp. 307–324). New York: Rockefeller Univ. Press.

Spielman, A. I., Mody, I., Brand, J. G., Whitney, G., MacDonald, J. F., & Salter, M. W. (1989). A method for isolating and patch-clamping single mammalian taste receptor cells. *Brain Research, 503,* 326–329.

Spielman, A. I., Nagai, H., Sunavala, G., Dasso, M., Breer, H., Boekhoff, I., Huque, T., Whitney, G., & Brand, J. G. (1996). Rapid kinetics of second messenger production in bitter taste. *American Journal of Physiology, 270,* C926–C931.

Streim, B. J., Naim, M., & Lindemann, B. (1991). Generation of cyclic AMP in taste buds of the rat circumvallate papilla in response to sucrose. *Cellular Physiology and Biochemistry, 1,* 46–54.

Streim, B. J., Pace, U., Zehavi, U., Naim, M., & Lancet, D. (1989). Sweet tastants stimulate adenylate cyclase coupled to GTP-binding protein in rat tongue membranes. *Biochemical Journal, 260,* 121–126.

Tabata, S., Crowley, H. H., Bottger, B., Finger, T. E., Margolskee, R. F., & Kinnamon, J. C. (1995). Immunoelectron microscopic analysis of gustducin in taste cells. *Chemical Senses, 20,* 296.

Tal, M., Ammar, D. A., Karpuj, M., Krizhanovsky, V., Naim, M., & Thompson, D. A. (1995). A novel putative neuropeptide receptor expressed in neural tissue including sensory epithelia. *Biochemical and Biophysical Research Communications, 209,* 752–759.

Teeter, J. H., Kumazawa, T., Brand, J. G., Kalinoski, D. L., Honda, E., & Smutzer, G. (1992). Amino acid receptor channels in taste cells. In D. P. Corey & S. D. Roper (Eds.) *Sensory transduction* (pp. 291–306). New York: Rockefeller Univ. Press.

Teeter, J. H., Sugimoto, K., & Brand, J. G. (1989). Ionic currents in taste cells and reconstituted taste epithelial membranes. In J. G. Brand, J. H. Teeter, R. H. Cagan, & M. R. Kare (Eds.) *Chemical senses. Volume 1. Receptor events and transduction in taste and olfaction* (pp. 151–170). New York: Dekker.

Tennissen, A. M., & McCutcheon, N. B. (1996). Anterior tongue stimulation with amiloride suppresses NaCl saltiness, but not citric acid sourness in humans. *Chemical Senses, 21,* 113–120.

Tonosaki, K. (1990). Taste transduction mechanisms. *Neuroscience Research, 12*(Suppl.), S63–S72.

Tonosaki, K., & Funakoshi, M. (1988). Cyclic nucleotides may mediate taste transduction. *Nature, 331,* 354–356.

Torii, K., & Cagan, R. H. (1980). Biochemical studies of taste sensation. IX. Enhancement of L-[3H]glutamate binding to bovine taste papillae by 5'-ribonucleotides. *Biochimica et Biophysica Acta, 627,* 313–323.

Walters, D. E., Orthoefer, F. T., & DuBois, G. E. (Eds.) (1991). *Sweeteners: Discovery, molecular design, and chemoreception.* Washington, DC: American Chemical Society.

Wong, G. T., Gannon, K. S., & Margolskee, R. F. (1996). Transduction of bitter and sweet taste by gustducin. *Nature, 381,* 796–800.

Yamaguchi, S. (1967). The synergistic taste effect of monosodium glutamate and disodium 5′-inosinate. *Journal of Food Science, 32,* 473–478.

Yamaguchi, S., Yoshikawa, T., Ikeda, S., & Ninomiya, T. (1971). Measurement of the relative taste intensity of some L-α-amino acids and 5′-nucleotides. *Journal of Food Science, 36,* 846–849.

Ye, Q., Heck, G. L., & DeSimone, J. A. (1991). The anion paradox in sodium taste reception: Resolution by voltage-clamp studies. *Science, 254,* 724–726.

Yokomukai, Y., Breslin, P. A. S., Cowart, B. J., & Beauchamp, G. K. (1994). Sensitivity to the bitterness of iso-α-acids: The effects of age and interactions with NaCl. *Chemical Senses, 19,* 577.

CHAPTER 2

The Neural Code and Integrative Processes of Taste

David V. Smith
Mark B. Vogt

I. INTRODUCTION

The perception of saltiness, sweetness, sourness, or bitterness emerges from neural activity within the central nervous system. Humans use these psychological concepts to describe the sensations arising from stimulation of gustatory receptors by a variety of chemical stimuli. The information necessary for these perceptions is carried to the brain by the activity in peripheral taste nerves. Response profiles of peripheral gustatory nerve fibers reflect the way in which the sensitivities of taste receptor cells are distributed among these first-order neurons. Transduction of specific chemical stimuli (e.g., sodium ions, protons, sugars, or alkaloids) by taste receptors gives rise to activity in several types of afferent nerve fibers. Understanding the neural coding of taste information begins with knowledge about how chemical sensitivities, represented by specific transduction mechanisms, are distributed and organized among peripheral and central gustatory neurons. The role of an individual taste fiber or central neuron in the coding of taste quality or other sensory parameters must be considered in the context of the multiple sensitivities of these cells. Chemical stimulation of taste receptors produces taste sensations, but also provides input that is critical for diverse somatic and visceral responses related to food ingestion and rejection. Viewing taste as the oral component of the visceral afferent system provides an important perspective on the involvement of individual gustatory neurons in the control of taste-mediated behaviors.

II. TASTE SYSTEM ANATOMY

A. Taste Bud Populations

Taste receptors in mammals are distributed on the tongue and throughout the oral, pharyngeal, and laryngeal epithelium within several subpopulations of taste buds that are innervated by one of three cranial nerves (Miller & Bartoshuk, 1991). In the hamster, for example, about 18% of the taste buds are located in the fungiform papillae on the anterior portion of the tongue, 32% are within the foliate papillae on the posterior sides of the tongue, 23% in the single midline vallate papilla on the posterior tongue, about 14% on the palate, distributed between the nasoincisive papillae (2%) and the soft palate (12%), and about 10% on the laryngeal surface of the epiglottis and the aryepiglottal folds (Miller & Smith, 1984). There are also a small number of taste buds within the sublingual organ, the buccal walls, the nasopharynx, and the upper reaches of the esophagus. Similar distributions are seen in the rat (Hosley, Hughes, & Oakley, 1987; Miller, 1977; Miller & Spangler, 1982) and other mammalian species that have been examined, including humans (Bradley, 1972; Bradley, Cheal, & Kim, 1980; Elliott, 1937; Khaisman, 1976; Miller & Bartoshuk, 1991; Nilsson, 1979).

B. Innervation of Taste Buds

1. Facial (VIIth) Nerve

Taste buds in the fungiform papillae on the anterior portion of the tongue and in the more rostral of the foliate papillae on the sides of the tongue are innervated by the chorda tympani (CT) branch of the facial (VIIth) nerve (Fish, Malone, & Richter, 1944; Oakley, 1988; Whiteside, 1927; Zalewski, 1969). Axons of the CT travel to the anterior tongue along with those of the lingual nerve (a branch of the mandibular division of the Vth nerve), which carries somatosensory innervation from the same area. The CT also carries preganglionic parasympathetic efferent fibers to the submandibular ganglion, which supplies the submandibular and sublingual salivary glands (Moore, 1992). The greater superficial petrosal (GSP) branch of the VIIth cranial nerve innervates taste buds on the soft palate via the lesser palatine nerve and in the nasoincisor ducts via the nasopalatine nerve (Cleaton-Jones, 1971; Miller, 1977; Miller & Spangler, 1982). The GSP also contains preganglionic parasympathetic fibers that travel to the pterygopalatine ganglion, where they synapse with postganglionic cells innervating the mucous glands of the hard and soft palate (Moore, 1992). The neurons giving rise to the gustatory fibers of the CT and GSP are located within the geniculate ganglion of the facial nerve. These two branches of the VIIth nerve carry gustatory information to the rostral pole of the nucleus of the solitary tract (NST), where their afferent terminations are largely coextensive (Contreras, Beckstead, & Norgren, 1982; Hamilton & Norgren, 1984; Travers, 1993).

2. Glossopharyngeal (IXth) Nerve

The vallate and foliate papillae on the posterior tongue contain taste buds inner-vated by the lingual-tonsillar branch of the glossopharyngeal (IXth) nerve (Guth, 1957; Oakley, 1970, 1988; Whiteside, 1927), which also supplies general somato-sensory fibers to the posterior one-third of the tongue. There is also some evidence that taste buds within the nasopharynx are innervated by the IXth cranial nerve (Travers & Nicklas, 1990). In rats and hamsters, over half of all the taste buds are distributed within the vallate and foliate papillae (Hosley et al., 1987; Miller & Smith, 1984, 1989). The IXth nerve carries preganglionic parasympathetic fibers to Remak's ganglion, from which postganglionic fibers supply the lingual salivary glands (Von Ebner's and Weber's glands). The parotid gland is supplied by pregan-glionic IXth nerve fibers that travel to the otic ganglion in the lesser petrosal nerve and then to the gland via postganglionic fibers in the great auricular and auriculo-temporal nerves (Moore, 1992). Afferent gustatory fibers of the IXth nerve, the cell bodies of which lie in the petrosal ganglion, project into the medulla and terminate within the NST somewhat caudal to, but overlapping with, the termination of the VIIth nerve (Contreras et al., 1982; Hamilton & Norgren, 1984; Hanamori & Smith, 1989; Sweazey & Bradley, 1986).

3. Vagus (Xth) Nerve

Taste buds distributed on the laryngeal surface of the epiglottis, on the aryepiglottal folds, and in the upper reaches of the esophagus are innervated by the internal branch of the superior laryngeal nerve (SLN), which is a branch of the vagus (Xth) nerve (Feindel, 1956; Khaisman, 1976). This nerve also carries somatosensory in-nervation from the supraglottic portion of the laryngeal mucosa (Moore, 1992). Chemosensitive fibers of the SLN, whose cell bodies lie within the nodose ganglion of the vagus nerve, project into the NST caudal to those of the VIIth and IXth cranial nerves (Contreras et al., 1982; Hamilton & Norgren, 1984; Hanamori & Smith, 1986, 1989; Sweazey & Bradley, 1986). Epiglottal taste buds are found in all mammalian species that have been studied and in the hamster are as numerous as those on the soft palate and almost as abundant as those in the fungiform papillae on the anterior tongue (Belecky & Smith, 1990; Miller & Smith, 1984).

C. Central Taste Pathways

The gustatory system is organized in a system of bilateral nuclei and tracts extending from caudal to rostral levels of the brain (see Travers, 1993, for a detailed review). Afferent fibers of the VIIth, IXth, and Xth cranial nerves carry gustatory informa-tion to the NST, the medullary relay for the gustatory and visceral afferent systems. These fibers terminate in the rostral pole of the NST (VIIth nerve) and at inter-mediate (IXth nerve) and more caudal (Xth nerve) levels. There is some overlap in their terminal fields within the NST, which lie predominantly rostral to the projection of general visceral afferent fibers of the vagus (Altschuler, Bao, Bieger, Hopkins, & Miselis, 1989; Contreras et al., 1982; Hamilton & Norgren, 1984; Hanamori & Smith, 1986, 1989). A schematic diagram of the major ascending projections within the gustatory pathway is shown in Figure 1. From the NST, ascending fibers project in most species to third-order cells within the parabrachial nuclei (PbN) of the pons (Halsell, 1992; Norgren, 1985), more or less parallel to the projection of general visceral sensation from the caudal NST (Herbert, Moga, & Saper, 1990). A thalamocortical projection arises from the PbN to carry taste information to the parvicellular portion of the ventroposteromedial nucleus of the thalamus (VPMpc) and on to gustatory neocortex (GN). In primates, taste fibers bypass the pontine relay and project directly to the VPMpc (Beckstead, Morse, & Norgren, 1980; see Pritchard, 1991 for review), as indicated by the dashed line in Figure 1. From the thalamus, taste information passes to cells in the GN, located within the agranular insular cortex (Kosar, Grill, & Norgren, 1986a, 1986b; Norgren & Wolf, 1975).

Arising in parallel with the thalamocortical projection is a limbic forebrain projection that carries gustatory afferent information into brain areas involved in feeding and autonomic regulation, including the lateral hypothalamus (H), central amygdala (A), and the bed nucleus of the stria terminalis (Halsell, 1992; Norgren, 1974, 1976, 1985; Norgren & Leonard, 1973). Descending projections within the gustatory system (not shown in Figure 1) have been demonstrated between the

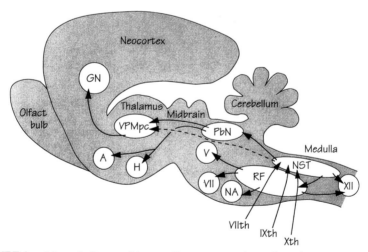

FIGURE 1   Schematic diagram of the ascending gustatory pathway; descending projections are not shown. Connections of the rodent gustatory system within the CNS are shown by solid lines; the projection from NST to VPMpc in primates is indicated by a dashed line. Abbreviations: NST, nucleus of the solitary tract; PbN, parabrachial nuclei; VPMpc, venteroposteromedial nucleus (parvi cellularis) of the thalamus; GN, gustatory neocortex; A, amygdala; H, hypothalamus; NA, nucleus ambiguus; RF, reticular formation; V, VII, and XII, trigeminal, facial, and hypoglossal motor nuclei; VIIth, IXth, and Xth, axons of peripheral gustatory fibers in the facial, glossopharyngeal, and vagal cranial nerves.

insular cortex and ventral forebrain and the PbN and NST (Saper, 1982; Saper, Swanson, & Cowan, 1979; Shipley, 1982; Shipley & Sanders, 1982; Van der Kooy, Koda, McGinty, Gerfen, & Bloom, 1984). There are also numerous local connections among neurons within the NST and with cells of the oral, facial, and pharyngeal motor nuclei (V, VII, ambiguus, and XII), either directly (as with XII) or via interneurons in the reticular formation (Travers, 1985; Travers & Norgren, 1983). These hindbrain systems form the substrate for many taste-mediated somatic and visceral responses related to ingestion and rejection of tastants (Grill & Norgren, 1978a,b; Kawamura & Yamamoto, 1978; Mattes, 1987).

III. PHYSIOLOGY OF GUSTATORY AFFERENT NEURONS

Most neurophysiological studies of the gustatory system have focused on the neural representation of the gustatory qualities of salty, sour, sweet, and bitter. Single peripheral gustatory fibers and central neurons are broadly tuned across the four taste qualities and often respond to tactile and thermal stimuli as well. Taste fibers in the facial, glossopharyngeal, and vagal nerves respond differentially to taste stimuli. These diverse sensitivities, which reflect both the quality and the hedonic value of the stimuli, provide input to neural systems related to ingestive and protective responses.

A. What Is Being Coded?

The gustatory system extracts three types of information from chemical stimuli: quality, intensity, and hedonic value. The unique perception of taste quality is the defining feature of this sense. Most researchers agree that in the absence of olfactory cues much of gustatory experience can be described by the sweet, salty, sour, and bitter qualities (McBurney & Gent, 1979). However, considerable debate continues over whether some other qualities (e.g., umami) should be included or whether any qualities are unique categories of taste perception (Erickson, 1977; McBurney, 1974; Schiffmann & Erickson, 1980; Yamaguchi, 1979). Intensity is a dimension common to all sensory systems, reflecting the magnitude of the evoked sensation. Hedonic value, the perceived pleasantness or unpleasantness of a taste sensation, is a response based on genetic, physiologic, and experiential factors as well as the characteristics of the stimulus. Taste is an inherently hedonic sense, relating strongly to motivated behavior (Pfaffmann, 1964; Stellar, 1977). Although the 3 dimensions of taste information can be assessed separately, they are not independent. For example, the perceived qualities of some taste stimuli (e.g., NaCl or Na-saccharin) change with stimulus concentration (Bartoshuk et al., 1978; Dzendolet & Meiselman, 1967). Similarly, the hedonic value of a stimulus is largely determined by its quality and intensity. Many omnivorous mammals share concentration-dependent preferences for substances humans describe as tasting sweet or salty and aversions to substances humans term sour or bitter (Beebe-Center, Black, Hoffman, & Wade,

1948; Berridge & Grill, 1984; Carpenter, 1956; Pfaffmann, 1964; Richter & Campbell, 1940). Presumably, these predispositions reflect evolutionary pressures related to the ingestional consequences (i.e., nutritional or toxic) of potential foods. The organization of the gustatory system is largely determined by genetic and developmental factors that configure the system and produce many "hardwired" functions. However, there are also mechanisms allowing experiential and physiologic factors to influence gustatory neural processing and perception. Species-specific predispositions regarding the hedonic value of a stimulus can be overcome via learned preferences or aversions (Bertino, Beauchamp, & Engelman, 1986; Grill, 1985; London, Snowdon, & Smithana, 1979; Ramirez, 1991; Rozin & Schiller, 1980) or in response to metabolic or pharmacologic manipulations (Parker, Maier, Rennie, & Crebolder, 1992; Richter, 1956). Even the perception of intensity and quality can be influenced by experience, as demonstrated by the performance of "expert tasters" and by improvements in the capacity of naive subjects to discriminate small differences during psychophysical testing. There is also evidence that experiential factors and physiologic state can influence the neural processing of gustatory information (Chang & Scott, 1984; Giza & Scott, 1983, 1987; Giza, Scott, & Vanderweele, 1992; Glenn & Erickson, 1976).

Thus gustatory afferent input provides at least three types of information that are interrelated in complex ways. How taste intensity, quality, and hedonic value are represented in the nervous system is the problem of gustatory neural coding. Historically, neurobiological research has approached the coding of these three dimensions as separate issues, with little attention to how they interrelate. In this chapter, we summarize current views of coding in the gustatory system. At the outset, we note two features of the taste system that have important implications for theories of gustatory coding. First, individual gustatory neurons, both peripheral and central, typically respond to stimuli representing several different taste qualities (Erickson, Doetsch, & Marshall, 1965; Frank, 1991; Frank, Bieber, & Smith, 1988; Hanamori, Miller, & Smith, 1988; Ogawa, Sato, & Yamashita, 1968; Pfaffmann, 1941, 1955; Scott & Giza, 1990; Smith, Van Buskirk, Travers, & Bieber, 1983a, 1983b; Travers & Smith, 1979; Van Buskirk & Smith, 1981). This is illustrated in Figure 4, which shows that single hamster CT fibers can respond robustly to two or even three different types of stimuli. Hamster taste neurons are more broadly tuned in the NST and PbN than in the CT nerve (Smith & Travers, 1979; Travers & Smith, 1979; Van Buskirk & Smith, 1981), and even in the periphery, taste fibers become more broadly tuned as stimulus concentration increases (Hanamori et al., 1988). Because the responses of taste fibers can be modulated by both quality and intensity, the response of any one neuron alone is entirely ambiguous with respect to either parameter (Pfaffmann, 1955, 1959; Travers & Smith, 1979; Van Buskirk & Smith, 1981). In addition, gustatory neurons are often responsive to thermal (Ogawa et al., 1968; Travers & Smith, 1984) and tactile stimuli (Hanamori, Ishiko, & Smith, 1987; Pfaffmann, Erickson, Frommer, & Halpern, 1961; Travers, 1993). This multimodal sensitivity is not entirely unexpected because taste is only one aspect of oral sensation and both temperature and touch have been shown to influence taste perception (Bartoshuk, Rennert, Rodin, & Stevens, 1982; Green & Frankmann, 1987; Moskowitz & Arabie, 1970). Thus impulse traffic in a single neuron may be related to several stimulus modalities, making the unambiguous interpretation of that signal impossible without comparing it to activity in other cells (Crick, 1979; Erickson, 1970, 1982, 1984). Therefore, in thinking about how sensory information is coded in the gustatory system, it is important to remember that cells at all levels of the pathway are broadly responsive to stimuli that vary in perceptual quality, are more broadly responsive at high than at low intensities, and are often sensitive to other modalities, such as touch and temperature.

The second important feature of the taste system is that, unlike most other sensory systems, there is no distinct topographic arrangement in its central neural organization (see Travers, 1993, for a thorough review of this issue). This lack of topography is not surprising because neither the molecules that constitute taste stimuli nor the perceptions of sweet, salty, sour, or bitter correspond to any continuous dimension of matter, space, or energy. There is some segregation of peripheral nerve terminations within the NST, and the taste bud subpopulations innervated by these nerves display somewhat different patterns of chemical sensitivity. This differential sensitivity results in some rough topographic differences in responsiveness within the NST (e.g., Dickman & Smith, 1989; Halpern, 1967; Halpern & Nelson, 1965). The segregation of peripheral nerve projections continues throughout the gustatory pathway to the cortex, where there are separate terminal fields for VIIth and IXth nerve inputs (Yamamoto, 1984). This segregation has been proposed to be important in the neural coding of taste quality (Yamamoto & Yuyama, 1987; Yamamoto, Yuyama, Kato, & Kawamura, 1985). However, central taste neurons at all levels have broadly tuned response characteristics (see Travers, 1993) and many have receptive fields in two or more receptor populations (Travers & Norgren, 1991; Travers, Pfaffmann, & Norgren, 1986). At present there is little compelling evidence that taste quality is represented by a strict topographic code. Thus, theories of gustatory coding must accommodate the multiquality and multimodal sensitivity of individual neurons as well as the lack of a strong topographical organization.

The majority of this chapter will deal with taste quality coding, although the coding of intensity and hedonic value are briefly addressed here. It is generally assumed, if not explicitly stated, that gustatory stimulus intensity is coded by neural impulse frequency. This assumption follows from observations that increases in stimulus concentration are associated with increases in perceived taste intensity and in neural response frequency (e.g., Ganchrow & Erickson, 1970; Scott & Perrotto, 1980). All neurons responsive to taste stimuli show some modulation by stimulus concentration; there is no evidence that only a specific subset of cells is responsible for coding stimulus intensity. Unlike gustatory intensity and quality, the issue of hedonic coding has not been systematically addressed in neurobiological studies of the taste system. This is probably because hedonic value is not independent of either quality or intensity and can be modified by both experience and physiologic state. Throughout this chapter we will attempt to relate the literature on gustatory physiology to the underlying hedonic dimension and to the appetitive and aversive responses that reflect hedonic value. This hedonic dimension is critically important in the control of many taste-mediated responses related to food ingestion and rejection.

B. Electrophysiology of Peripheral Taste Fibers

The various subpopulations of taste buds are differentially sensitive to taste quality, resulting in differences among the gustatory nerves in the sensitivities of the afferent axons. Facial nerve fibers are relatively more responsive to preferred stimuli (sugars and salt), whereas taste axons of the glossopharyngeal nerve respond preferentially to aversive stimuli (acids and quinine). Vagal chemosensory fibers are responsive to stimuli that deviate from the normal pH and ionic milieu of the larynx.

1. Facial Taste Fibers

The response properties of gustatory fibers in the CT nerve, which innervates taste buds on the anterior part of the tongue, have been characterized in several mammalian species (Frank, 1973; Fishman, 1957; Ogawa et al., 1968; Pfaffmann, 1974; Pfaffmann, Frank, Bartoshuk, & Snell, 1976). Fibers of the hamster CT are differentially responsive to a wide array of gustatory stimuli (Frank, 1973; Frank et al., 1988), and input to brain stem gustatory nuclei from the anterior tongue of the hamster is sufficient to discriminate among several classes of stimuli (Smith et al., 1983b) that are behaviorally discriminable by hamsters (Frank & Nowlis, 1989; Nowlis & Frank, 1981; Smith, Travers, & Van Buskirk, 1979).

Since the earliest electrophysiological studies of single taste fibers (Pfaffmann, 1941, 1955), it has been recognized that individual fibers of the CT nerve are responsive to stimuli representing more than one of the classical four taste qualities (salty, sour, sweet, and bitter). Nevertheless, fibers within the CT nerve of the hamster (Frank, 1973), and of other species as well (Frank, Contreras, & Hettinger, 1983; Pfaffmann et al., 1976), can be grouped into classes based on their relative sensitivities to four taste stimuli. When the stimuli 0.1 M sucrose, 0.03 M NaCl, 3 mM HCl, and 1 mM quinine-HCl (QHCl) are ordered hedonically along the abscissa from most to least preferred, the response spectra of hamster CT fibers show a single peak as sucrose-, NaCl-, or HCl-best fibers (Frank, 1973). Few fibers ($< 1\%$) most responsive to this concentration of QHCl are found in the CT nerve. Thus taste fibers of the hamster CT nerve have an organization to their sensitivities; there are three neuron classes defined by their sensitivities to four basic stimuli applied to the anterior portion of the tongue: sucrose-, NaCl-, and HCl-best fibers. This kind of best-stimulus classification has been used extensively in subsequent studies of the hamster gustatory system (Frank et al., 1988; Smith & Hanamori, 1991; Smith et al., 1983a, 1983b; Travers & Smith, 1979; Van Buskirk & Smith, 1981) and that of other species (Frank, 1991; Frank et al., 1983; Pfaffmann et al., 1976).

The appropriateness of this best-stimulus classification of CT fibers was examined further by analyzing the responses of 40 CT fibers to an array of 13 stimuli (Frank et al., 1988). The response profiles of these 40 fibers are shown in Figure 2, which allows a comparison among them. Within this figure, the fibers are arranged according to their best stimulus, with *fibers 1– 10* being sucrose-best, *fibers 11– 32* NaCl-best, and *fibers 33– 40* HCl-best. The hatched bars in Figure 2 represent responses to four prototypical stimuli: sucrose (sweet), NaCl (salty), HCl (sour), and QHCl (bitter). The solid bars represent responses to other stimuli, shown along the abscissa. Although the response profiles within a best-stimulus group are not identical, inspection of Figure 2 gives the impression that there are essential similarities among profiles of fibers within a group and striking differences between profiles in different groups.

The response profiles shown in Figure 2 were subjected to a hierarchical cluster analysis, which addressed whether it is reasonable to assume that these profiles were sampled from distinct subpopulations rather than from a single population (Bieber & Smith, 1986). In this analysis, the most similar pairs of profiles are clustered together first, followed by the clustering of pairs of profiles that are more dissimilar. The resulting dendrogram depicting the hierarchical arrangement of this clustering is shown in Figure 3. At the right of the figure, the fiber numbers are indicated as they are specified in Figure 2. The analysis segregated the CT fiber profiles into three major clusters, members of which are connected by solid lines in Figure 3. This conclusion is based on a regular, stepwise increase in the intercluster distance as the linking proceeds, until, in moving from three clusters to two, a dramatic increase in the intercluster distance occurs. The fiber classes defined by the cluster analysis are labeled "S", "H" and "N" in Figure 3, which correspond, with one exception, to the sucrose-, HCl-, and NaCl-best fibers, respectively. Thus, within the hamster CT nerve, classification of fibers by using four prototypical stimuli produces the same classification as a hierarchical cluster analysis based on response profiles across 13 stimuli. Similar classes emerged from hierarchical clustering of second- and third-order brain stem neurons in the hamster when as few as 3 (McPheeters et al., 1990) or as many as 18 (Smith et al., 1983a) stimuli were applied to the anterior portion of the tongue. The stimulus array must, however, include at least one example of three stimulus classes: (1) sweeteners, (2) sodium salts, and (3) nonsodium salts and acids (see later).

When this array of 13 stimuli was applied to the hamster's fungiform papillae, the responses of the CT fibers reflected the perceptual similarities and differences among the stimuli, as determined by studies of behavioral generalization among taste stimuli by hamsters (Frank & Nowlis, 1989; Nowlis & Frank, 1981; Smith et al., 1979). As previously demonstrated for rat CT fibers (Erickson, 1963; Erickson et al., 1965), stimuli with similar taste quality produce patterns of activity across

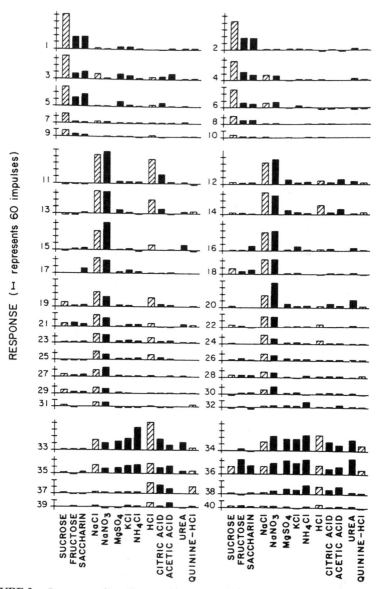

FIGURE 2   Response profiles of hamster chorda tympani nerve fibers. The numbers to the left of each profile identify the fibers. The left-hand column shows profiles for odd-numbered fibers and the right-hand column shows profiles for even-numbered fibers. Fibers 1–10 are sucrose-best, fibers 11–32 are NaCl-best, and fibers 33–40 are HCl-best. Test stimuli are listed along the abscissa, beneath bars whose heights represent response rates. The response (nerve impulses) rates for 5 sec are indicated. The long horizontal lines from which the bars project represent a rate of zero. Each short horizontal line sequentially crossing the profile's ordinate represents an additional rate increase of 60 impulses above the spontaneous rate. Response rates that are lower than the spontaneous rate are seen as bars extending

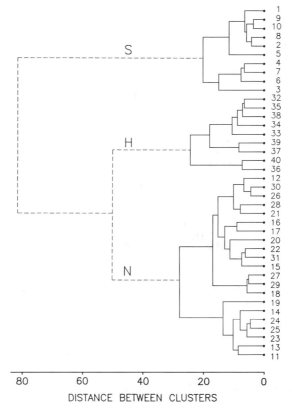

DISTANCE BETWEEN CLUSTERS

FIGURE 3   Cluster analysis of hamster chorda tympani fiber response profiles. The parallel horizontal lines of the dendrogram represent profiles, or groups of profiles, of fibers indicated at the right and
numbered as in Figure 2. The major profile clusters, sucrose-best, HCl-best, and NaCl-best (S, H, N),
are identified to the left of the defining vertical lines. The distances between profiles (or groups of
profiles) are obtained by projecting the vertical lines to the distance scale along the abscissa (Modified
from a figure in Frank, Bieber, & Smith, 1988). Reproduced from *The Journal of General Physiology,* 1988,
91, 861–896, by copyright permission of the Rockefeller University Press.

hamster CT fibers that are highly correlated. The across-fiber patterns for 13 stimuli
are depicted in Figure 4, which shows that stimuli with similar tastes produce similar
patterns of activity. For example, the correlation between the patterns produced by
sucrose and Na-saccharin was +0.89; between NaCl and NaNO$_3$, +0.93; and between KCl and NH$_4$Cl, +0.91. Within this stimulus array, three distinctly different

below the horizontal zero line. The hatched bars represent responses to prototypical stimuli (sucrose,
NaCl, HCl, and QHCl) and the filled bars represent responses to other stimuli (Frank, Bieber, & Smith,
1988). Reproduced from *The Journal of General Physiology,* 1988, *91,* 861–896, by copyright permission
of the Rockefeller University Press.

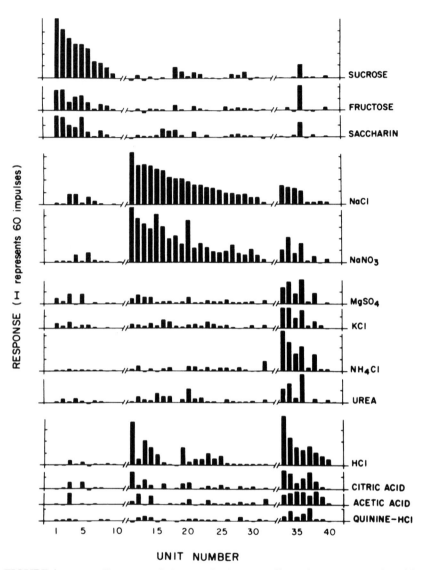

UNIT NUMBER

FIGURE 4   Across-fiber patterns for hamster chorda tympani fibers. The response rates elicited for 5 sec in chorda tympani fibers 1–40 (Figure 2) are represented by consecutive filled bars from left to right. Each row depicts the pattern elicited by the stimulus indicated to the right of the horizontal line representing the zero-response rate. These horizontal lines are broken twice with two short parallel diagonals at the transitions between sucrose-best and NaCl-best (between fibers 10 and 11) and NaCl-best and HCl-best (between fibers 32 and 33) fibers. Fiber numbers are indicated along the abscissa. The short horizontal lines parallel to the zero-rate line interrupting the ordinates indicate successive increases in the response rate of 60 nerve impulses. The bars extending below the zero line indicate response rates lower than the spontaneous rate (Frank, Bieber, & Smith, 1988). Reproduced from *The Journal of General Physiology*, 1988, *91*, 861–896, by copyright permission of the Rockefeller University Press.

patterns are seen. One is elicited by sucrose, fructose, and Na-saccharin. A second pattern is elicited by NaCl and NaNO₃. A third, more variable pattern is evoked by the remaining stimuli, including the acids, nonsodium salts, urea, and QHCl. Within these patterns, the most responsive neurons for a particular group of stimuli tend to fall within one of the best-stimulus classes of fibers. For example, for the patterns evoked by sucrose, fructose, and Na-saccharin, the sucrose-best fibers (*fibers 1–10*) are the most responsive. For the sodium salts, the NaCl-best fibers (*fibers 11–32*) are the most responsive; however, the HCl-best fibers are often quite responsive as well. For the nonsodium salts and acids the HCl-best fibers (*fibers 33–40*) are most responsive, although HCl also activates NaCl-best fibers. Thus within the activity elicited in the CT nerve, the responses of particular sets of fibers (S, N, or H fibers) typically dominate the patterns evoked by particular sets of stimuli (sweet tasting, sodium salts, or nonsodium salts and acids).

The organization of sensitivities within the CT nerve suggests that at least three kinds of receptor mechanisms provide input to fibers of this nerve. These transduction mechanisms define the spectrum of chemicals to which sucrose-, NaCl-, and HCl-best nerve fibers respond. There is a growing body of literature suggesting that one receptor mechanism for sodium salts is an amiloride-blockable sodium channel on the apical membrane of taste receptor cells (Avenet & Lindemann, 1988; DeSimone et al., 1984; Formaker & Hill, 1988; Heck et al., 1984; Kinnamon, 1988; Simon, Robb, & Schiffman, 1988). Application of amiloride to the tongue reduces the response to NaCl in single CT nerve fibers of hamsters and rats (Hettinger & Frank, 1990; Hill, 1987; Ninomiya & Funakoshi, 1988) and responses of the whole CT nerve in several mammalian species (Brand, Teeter, & Silver, 1985; DeSimone & Ferrell, 1985; Heck et al., 1984; Hellekant, DuBois, Roberts, & van der Wel, 1988; Herness, 1987; Hettinger & Frank, 1990). These data imply that taste stimulation by sodium salts may involve entry of the stimulus into taste receptor cells through an amiloride-sensitive epithelial ion channel (DeSimone et al., 1984). A second transduction pathway appears to be involved in the transduction of NaCl, but not of sodium salts with large cations, such as Na-gluconate (Ye, Heck, & DeSimone, 1993); this mechanism involves ion channels on the basolateral membranes of taste receptor cells that can be accessed via diffusion through a paracellular pathway.

Taste buds in the nasoincisor ducts, along the "Geschmacksstreifen" (in the rat), and on the soft palate are innervated by fibers of the GSP nerve (Miller, 1977; Miller & Spangler, 1982). Although it has not yet been possible to analyze the responses of single gustatory fibers of the GSP, recent studies have examined the responsiveness of the whole GSP nerve in the rat (Nejad, 1986) and in the hamster (Harada & Smith, 1992). In addition, rat NST single-neuron responses to stimulation of the nasoincisor ducts and soft palate have been examined (Travers & Norgren, 1991; Travers et al., 1986). In general, the palate of both rats and hamsters is more responsive to sweet-tasting stimuli than is the anterior tongue, although this difference is more striking in the rat. The CT nerve of the rat is relatively insensitive to sucrose and to other compounds that are sweet to humans (Frank et al., 1983; Pfaffmann, 1955). This has traditionally been puzzling to taste physiologists, because rats show an avid preference for sweet-tasting compounds (Beebe-Center et al., 1948; Davis, 1973; Richter & Campbell, 1940). However, both the palate (Nejad, 1986; Travers & Norgren, 1991; Travers et al., 1986) and the posterior tongue (Boudreau, Do, Sivakumar, Oravec, & Rodriquez, 1987; Frank, 1991) of the rat are more sensitive than the anterior tongue to sucrose and other sweet stimuli. Even in the hamster, whose CT nerve is quite responsive to sucrose, the GSP is more responsive to sweet-tasting stimuli (Harada & Smith, 1992).

2. Glossopharyngeal Taste Fibers

Studies of the responsiveness of the whole glossopharyngeal (IXth) nerve in several mammalian species have suggested that its sensitivities to gustatory stimuli are different from those of the CT (Mistretta & Bradley, 1983; Oakley, 1967; Oakley, Jones, & Kaliszewski, 1979; Shingai & Beidler, 1985; Yamada, 1966, 1967). For example, the IXth nerve of the rat contains many fibers that are tuned to respond best to QHCl or to sucrose, which are poor stimuli for the rat CT nerve (Frank, 1991). Responses of cells in the rat petrosal ganglion, which contains the cell bodies of the IXth nerve afferent fibers, also suggest a greater responsiveness to QHCl and sucrose than is seen in the CT nerve (Boudreau et al., 1987).

Fibers of the IXth cranial nerve of the hamster are predominantly responsive to HCl and QHCl and show much less responsiveness to sucrose and NaCl over a wide range of concentrations (Hanamori et al., 1988). The mean responses of 56 hamster IXth nerve fibers innervating the foliate papillae and of 27 fibers innervating the vallate papilla to five concentrations each of sucrose, NaCl, HCl, and QHCl are shown in Figure 5. A single IXth nerve fiber innervated either the vallate or the foliate papillae, but never both; flowing solutions into the trenches of these separate papillae never stimulated the same fiber; this is also true of rat IXth nerve fibers (Frank, 1991). Sensitivity to QHCl is greater in the IXth nerve than in the CT, but the threshold for QHCl in hamster IXth nerve fibers is relatively high (between 1 and 3 mM). This threshold, however, corresponds to the point where a clear aversion for QHCl emerges in two-bottle preference experiments on hamsters (Carpenter, 1956). The stimulus concentrations that were half-maximal for the CT nerve (0.1 M sucrose, 0.03 M NaCl, 3 mM HCl, and 1 mM QHCl) were not very effective stimuli for the IXth nerve, which in itself suggests that the taste receptors of the anterior and posterior tongue are different. The response measure employed to examine the organization of IXth nerve sensitivities was the sum of the responses to all five concentrations (total response), which is proportional to the area under the concentration–response function (shaded in Figure 5). This measure ("best" total response) likely gives a more accurate characterization of the sensitivities of a neuron than the response to a single concentration, because any best-stimulus classification depends on the choice of concentrations (Frank et al., 1983; Maes, 1985; Travers & Smith, 1979; Van Buskirk & Smith, 1981). Classification of IXth nerve

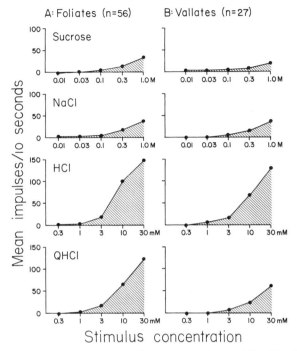

FIGURE 5   Mean concentration–response functions (nerve impulses/10 sec) for 56 hamster IXth nerve fibers innervating the foliate papillae (A) and 27 IXth nerve fibers innervating the vallate papilla (B). Responses are shown across five concentrations. The shaded area is proportional to the sum of the responses across these concentrations (total response), which was used as a response measure for each fiber. Reprinted from Hanamori, Miller, and Smith (1988), with permission.

fibers using this total response measure resulted in 8 sucrose-, 4 NaCl-, 52 HCl-, and 19 QHCl-best fibers. Most units (> 90%) were similarly classified using responses to 0.3 M sucrose, 0.3 M NaCl, 10 mM HCl, and 10 mM QHCl, which are midrange for the IXth nerve of the hamster (see Figure 5).

Based on stimulation of the anterior tongue with a single midrange concentration of each of four stimuli, Frank (1973) classified 79 hamster CT fibers into one of four classes: sucrose-best ($n = 20$), NaCl-best ($n = 42$), HCl-best ($n = 17$), or QHCl-best ($n = 1$). Thus many hamster CT fibers respond best to sucrose or NaCl. On the other hand, most hamster IXth nerve fibers respond best to HCl or QHCl. A comparison of the mean response profiles of CT and IXth nerve fibers is shown in Figure 6, which demonstrates differences in both the numbers of fibers of different types and their relative responsiveness to the four stimuli. Fibers in the IXth nerve are mostly responsive to HCl or QHCl (Figure 6A), whereas those in the CT nerve respond predominantly to sucrose or NaCl (Figure 6B).

In several mammalian species there is relatively poor sensitivity to QHCl in the CT nerve (Frank, 1973; Frank et al., 1988; Ogawa et al., 1968; Pfaffmann et al.,

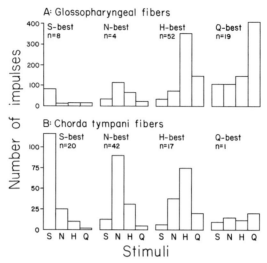

FIGURE 6   Mean response profiles for 83 fibers in the hamster IXth nerve (A) and 79 fibers in the hamster chorda tympani nerve (B). The number of impulses is the total number of impulses in 10 sec across five concentrations in A and number of impulses in 5 sec to a single midrange concentration in B. Reprinted from Hanamori, Miller, and Smith (1988), with permission.

1976). On the other hand, the IXth nerve is very responsive to QHCl in several species, including rats (Frank, 1991), gerbils (Oakley et al., 1979), mice (Shingai & Beidler, 1985), and sheep (Mistretta & Bradley, 1983). This QHCl sensitivity in the IXth nerve is also evident in fish and amphibians. The IXth and Xth nerves in catfish supply taste buds in the oral and pharyngeal epithelium and are similar to VIIth nerve fibers of the catfish in their response to amino acids; however, fibers in the IXth and Xth nerves are much more sensitive than VIIth nerve fibers to QHCl (Kanwal & Caprio, 1983). Similarly, IXth nerve fibers of several amphibian species show very low thresholds for quinine (Gordon & Caprio, 1985; Yoshii et al., 1981, 1982). Thus there is a general trend across vertebrate species for fibers in the IXth nerve to be more responsive to quinine than those in the VIIth nerve.

The hamster IXth nerve contains very few neurons that are specifically sensitive to sodium salts; only 4 of 83 fibers were classified as NaCl-best (Figure 6). Similarly, there were no fibers seen in the rat IXth nerve specifically tuned to NaCl (Frank, 1991), although NaCl was effective in driving units most responsive to acids and nonsodium salts. The ability of rats to show a previously learned discrimination between NaCl and KCl is completely disrupted if the CT nerves are cut, but this discrimination is not disturbed following bilateral transection of only the IXth nerve (Spector & Grill, 1992). Treatment of the tongue with amiloride during conditioning to avoid NaCl interferes with the ability of rats to make subsequent discriminations between sodium and nonsodium salts (Hill et al., 1990). Mixing NaCl with amiloride also increases the intake of normally nonpreferred NaCl solutions in hamsters (Hettinger & Frank, 1990). Amiloride has a specific effect on NaCl-best units in the CT nerve of these species (Hettinger & Frank, 1990; Ninomiya & Funakoshi, 1988). Consistent with the failure of IXth nerve transection to affect the behavior toward sodium salts and with the paucity of NaCl-best fibers in the IXth nerve is the fact that amiloride treatment does not affect the rat's IXth nerve response to NaCl (Formaker & Hill, 1991).

3. Vagal Chemosensory Fibers

Taste buds distributed on the laryngeal surface of the epiglottis, on the aryepiglottal folds, and in the upper reaches of the esophagus are innervated by the superior laryngeal nerve (SLN), which is a branch of the vagus (Xth) nerve (Feindel, 1956; Khaisman, 1976). The responsiveness of afferent fibers in the SLN to chemical stimulation of the larynx has been studied in several species. Electrophysiological studies that have directly compared the responsiveness of the SLN to that of other gustatory nerves have shown some striking differences in their sensitivities (Bradley, 1982; Bradley, Stedman, & Mistretta, 1983; Dickman & Smith, 1988; Shingai & Beidler, 1985; Smith & Hanamori, 1991; Stedman, Bradley, Mistretta, & Bradley, 1980; Sweazey & Bradley, 1988).

Responses of 65 single fibers in the SLN of the hamster were recorded following stimulation of the laryngeal epithelium with the same array of stimuli employed in the study of the IXth nerve (see Figure 5), plus distilled water (Smith & Hanamori, 1991). Because of the significant response to distilled water in these fibers, each stimulus except the NaCl concentration series was dissolved in 0.154 M NaCl, which produces a minimal discharge in SLN fibers. The mean responses to these 21 stimuli across these 65 fibers are shown in Figure 7, where it is evident that distilled water, HCl, and high concentrations of NaCl are much more effective

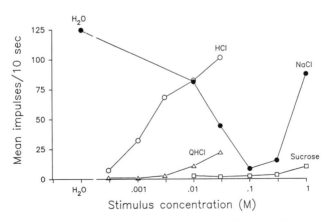

FIGURE 7 Mean concentration–response functions (impulses/10 sec) for 65 fibers in the hamster superior laryngeal nerve. All 65 fibers were stimulated with every stimulus concentration. Response to distilled H_2O is shown as the lower extreme of the NaCl series, which was dissolved in distilled H_2O. All other stimulus series were dissolved in 0.154 M NaCl, which was used as a rinse before and after each stimulus presentation. Reprinted from Smith and Hanamori (1991), with permission.

stimuli than is QHCl or sucrose. These fibers responded most to distilled water, and as the concentration of NaCl was increased (i.e., as water was "diluted") there was a decreasing response of SLN fibers, reaching a minimum around the adapting concentration of 0.154 M NaCl. Further increases in NaCl concentration produced increasing levels of response in these fibers. Similar concentration-dependent relationships have been shown for SLN fibers in other species (Boggs & Bartlett, 1982; Shingai, 1977, 1980; Storey, 1968; Storey & Johnson, 1975). Because increasing the concentration of NaCl but not of nonchloride sodium salts decreases the response to water, it has been proposed that the response to water is mediated through the outward movement of Cl^- ions through the receptor membrane (Boggs & Bartlett, 1982; Shingai, 1977). However, there are yet no direct electrophysiological studies on transduction by laryngeal chemoreceptors to support this hypothesis.

Responses of fibers in hamster SLN primarily reflect the input from three receptor mechanisms: water, acid, and sodium. Classifying these 65 SLN fibers according to which of five stimuli (distilled water, 1.0 M NaCl, 0.03 M HCl, 0.03 M QHCl, or 1.0 M sucrose) was the most effective stimulus resulted in 26 water-, 17 NaCl-, 20 HCl-, and 2 QHCl-best fibers. These concentrations were chosen for the best-stimulus classification because of the complex relationship in the water–NaCl function (Figure 7) and the general lack of responsiveness to QHCl and sucrose at all but the strongest concentrations. At these intensities, most fibers responded best to water, HCl, or NaCl rather than to QHCl or sucrose. Within these best-stimulus classes, fibers were not specifically tuned to their best stimulus. Water-best fibers (which also responded to hypotonic NaCl solutions) were also responsive to HCl and hypertonic NaCl. Similarly, NaCl-best fibers also responded moderately to water (and hypotonic NaCl) and to strong HCl. Fibers responding best to HCl also responded moderately to water and NaCl and somewhat to strong QHCl. These best-stimulus classes of cells were not as distinct from one another as those in the hamster CT or IXth nerve. The results of a hierarchical cluster analysis of the response profiles of these 65 SLN fibers indicated that there are not distinct clusters of chemosensitive fibers in the SLN (Smith & Hanamori, 1991). Rather, they appear to represent a continuum of response profiles, each somewhat different from the others.

C. Differential Gustatory Inputs to the Brain Stem

The data presented above for the four gustatory nerves of the hamster show that the various populations of taste buds differ in their sensitivities and contribute different kinds of afferent information to the brain stem. Although there are species differences in the distributions of some of these sensitivities, the general conclusion is that the various taste bud populations provide different kinds and amounts of gustatory information. It is likely that these variable inputs are important for different kinds of taste-mediated behavior.

The relative responsiveness of the hamster's four gustatory nerves (CT, IXth, GSP, and SLN) to 0.3 M sucrose, 0.3 M NaCl, 10 mM HCl, 10 mM QHCl, and

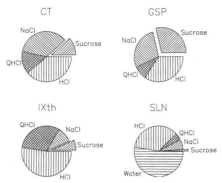

FIGURE 8   Proportional responses to 0.3 *M* sucrose, 0.3 *M* NaCl, 10 m*M* QHCl, 10 m*M* HCl, and distilled water in four gustatory nerves of the hamster. Data for the chorda tympani (CT) and greater superficial petrosal (GSP) were integrated tonic responses (Harada & Smith, 1992); those for the IXth and superior laryngeal nerve (SLN) are mean firing rates over a 10-sec response period in 83 IXth nerve fibers (Hanamori et al., 1988) and 65 SLN fibers (Smith & Hanamori, 1991). Each pie chart represents the response to each stimulus as a proportion of the sum of the responses of that nerve to all five stimuli. Reprinted from Harada and Smith (1992), *Chemical Senses 17,* by permission of Oxford University Press.

distilled water is shown in Figure 8. These particular concentrations were chosen for comparison because they were common to the several studies previously described (Hanamori et al., 1988; Harada & Smith, 1992; Smith & Hanamori, 1991). The figure depicts the response of each stimulus as a proportion of the total response to all five; thus the comparison among the nerves is relative. For the CT and GSP, the responses were the tonic integrated responses of the whole nerve (Harada & Smith, 1992). The responses for the IXth nerve and the SLN were mean impulses/ 10 sec across 83 IXth nerve fibers (Hanamori et al., 1988) and 65 SLN fibers (Smith & Hanamori, 1991). The differences in sucrose sensitivity are highlighted by the exploded portion of the pie charts. Receptors on the anterior tongue (CT) and palate (GSP) provide relatively more information about NaCl and sucrose than do receptors on the posterior tongue (IXth) or in the larynx (SLN). Sensitivity to HCl is relatively similar in every gustatory nerve; QHCl clearly has its greatest relative effect in the IXth nerve. Only the SLN of the hamster responds to distilled water; this response occurred after adaptation to 0.154 *M* NaCl, but neither the CT, the IXth, nor the GSP nerve responds to water after NaCl adaptation. Information arising from these various gustatory nerves projects into the nucleus of the solitary tract (NST) of the medulla, where appropriate connections provide for the reflexive control of ingestive and protective responses that are triggered by taste stimulation.

Behavioral studies have begun to define taste quality for a number of mammalian species, including rats and hamsters (Erickson, 1963; Frank & Nowlis, 1989; Hill et al., 1990; Morrison, 1967; Nachman, 1963; Nowlis & Frank, 1981; Nowlis et al., 1980; Smith & Theodore, 1984; Smith et al., 1979; Spector & Grill, 1988). These rodents easily discriminate among sucrose, NaCl, HCl, and QHCl; they group the tastes of other sugars and Na-saccharin with sucrose, other acids and some non-sodium salts such as NH_4Cl with HCl, other sodium salts with NaCl, and some bitter-tasting salts such as $MgSO_4$ with QHCl. Input from the fungiform papillae of the hamster is sufficient to allow neural and behavioral discrimination among sugars, sodium salts, and acids (Frank & Nowlis, 1989; Frank et al., 1988). In the rat, discrimination of sugars from other stimuli may depend primarily on input from the GSP nerve (Nejad, 1986; Spector, Schwartz, & Grill, 1990; Krimm, Nejad, Smith, Miller, & Beidler, 1987). However, neural activity in the CT nerve does not discriminate well between QHCl and the nonsodium salts and acids; this discrimination is much more dramatic in fibers of the IXth nerve (Frank, 1991). Thus the behavioral distinctions among stimuli with different taste qualities may rely somewhat on input from different cranial nerves. Fibers of the SLN innervating laryngeal chemoreceptors, however, do not appear to discriminate among stimuli with different taste qualities (Dickman & Smith, 1988). A multidimensional scaling analysis of the similarities among the across-fiber patterns generated by 20 stimuli, each at a single concentration, did not produce a separation between stimuli of different quality, as seen in the responses of CT (Frank et al., 1988) or IXth nerve (Frank, 1991) fibers. Rather, the predominant dimension separated stimuli that were excitatory for SLN fibers, such as KCl, acids, and urea, from those that were inhibitory, such as $CaCl_2$ and the sugars (Dickman & Smith, 1988). Thus chemosensitive fibers of the SLN appear to be suited to a role in airway protection, by signaling deviations from the normal pH and ionic milieu of the larynx (Boggs & Bartlett, 1982; Shingai & Shimada, 1976; Smith & Hanamori, 1991; Storey & Johnson, 1975) rather than in the discrimination among gustatory qualities.

IV. CODING OF TASTE QUALITY

There has been considerable controversy over whether taste quality is represented in the nervous system by activity in specific neural channels (called *labeled lines*), or by the relative activity across the population of responsive neurons (across-fiber patterns). The multiple sensitivity of taste-sensitive neurons makes a strict labeled-line hypothesis difficult to accept, although there is compelling evidence for taste neuron types based on similarities in their profiles of sensitivity. A consideration of the relationships between neuron types and across-fiber patterns suggests that taste quality coding can best be accounted for by the relative activity across neuron types.

A. Taste Coding Theories

The nature of the neural representation of taste quality has been vigorously debated for many years (Erickson, 1982, 1984, 1985; Frank et al., 1988; Pfaffmann, 1974; Pfaffmann et al., 1976; Scott & Giza, 1990; Scott & Plata-Salaman, 1991; Smith, 1985; Smith & Frank, 1993; Smith et al., 1979, 1983a, 1983b; Woolston & Erickson, 1979). Even the existence of four basic taste qualities is not universally accepted, with some authors insisting on additional qualities and others arguing that taste experience is a continuum on which the familiar qualities are merely arbitrary points (Erickson, 1977; McBurney, 1974; McBurney & Gent, 1979; Nowlis & Frank, 1981; Schiffmann & Erickson, 1980; Scott & Plata-Salaman, 1991).

Prior to the development of neurophysiological recording methods, a long tradition of human psychophysical research had provided considerable support for the notion that taste experience could be reduced to a few basic qualities, although not necessarily the traditional four (see Bartoshuk, 1971). This idea, combined with Meuller's doctrine of specific nerve energies, led to the expectation that the perception of taste quality would arise from the activation of one of a few neuron types, each coding a single taste quality. This strict "labeled line" theory was discounted by early neurophysiological recordings showing that peripheral taste fibers in several species were responsive to stimuli representing more than one taste quality (Pfaffmann, 1941, 1955). In response, an "across-fiber pattern" theory was proposed, which held that taste quality is coded by the relative activity across a population of neurons (Pfaffmann, 1959; Erickson, 1963). This theory accommodated the multiple sensitivity of taste fibers and required neither specific fiber types nor taste primaries. However, the persisting view of the importance of taste primaries led to a modification of the labeled-line theory, which proposed that taste quality is coded by the activity in a few "best-stimulus" channels, that is, by neurons that respond best, but not specifically, to one of the basic taste qualities (Pfaffmann, 1974; Pfaffmann et al., 1976).

Although each coding theory has its strengths, both strain to encompass the full range of data. For example, accumulating evidence suggests that there are indeed functional classes of neurons that correspond in some way to primary taste qualities (Boudreau & Alev, 1973; Contreras & Frank, 1979; Frank, 1973, 1991; Frank et al., 1988; Giza & Scott, 1991; Hanamori et al., 1988; Hettinger & Frank, 1990; Ninomiya & Funakoshi, 1988; Scott & Giza, 1990; Smith et al., 1983a; Smith & Frank, 1993). On the other hand, recent analyses show that no single class of neurons in isolation can discriminate between different taste qualities (Smith, 1985; Smith & Frank, 1993; Smith et al., 1983b). The neural coding problem essentially rests on whether the activity in a given taste fiber is an unambiguous representation of the quality of the stimulus applied to its receptors or whether this activity is meaningful only in the context of activity in other afferent fibers. In this chapter we review the relevant neurophysiological data that bear on this issue and suggest that taste quality is coded in the relative rates of activity across several neuron types (Smith, 1985; Smith & Frank, 1993).

1. Across-Fiber Patterns

It was the multiple sensitivity of fibers in the CT nerve that first led Pfaffmann (1941, 1955, 1959) to propose that taste quality is coded by the pattern of activity across taste fibers. With this coding hypothesis, taste quality remains invariant with increased intensity, even though any single neuron may increase its breadth of responsiveness. The pattern of activity generated across the entire array of taste neurons at a higher concentration is similar in shape, but varies in amplitude (Erickson, 1982, 1984, 1985; Erickson et al., 1965; Ganchrow & Erickson, 1970). The across-fiber patterns to an array of 13 stimuli were shown in Figure 4 for 40 hamster CT fibers. It is obvious in Figure 4 that stimuli with similar tastes, such as the sweeteners or the sodium salts, generate highly similar patterns of activity across these fibers. These similarities are typically measured by calculating the across-fiber correlation between pairs of stimuli (Erickson et al., 1965), although other indices have been proposed (DiLorenzo, 1989; Gill & Erickson, 1985). Several behavioral investigations have shown that stimuli that evoke highly correlated neural patterns are judged by experimental animals to have similar tastes (Erickson, 1963; Morrison, 1967; Nowlis & Frank, 1981; Nowlis et al., 1980; Smith et al., 1979; Wiggens et al., 1989). This across-fiber pattern view of quality coding makes the multiple sensitivity of gustatory neurons an essential part of the neural code for taste quality. This theoretical view stresses that the code for quality is given in the response of the entire population of cells (Erickson, 1968, 1977, 1982, 1984, 1985), placing little or no emphasis on the role of an individual neuron. Erickson (1968, 1974, 1982, 1984, 1985) has argued that such a coding mechanism could operate for many sensory systems, particularly for nontopographic modalities employing neurons that are broadly tuned across their stimulus array.

When the across-neuron correlations are calculated for an array of gustatory stimuli across ether peripheral or central neurons, stimuli with similar tastes correlate highly and those with different tastes correlate less. Almost every neurophysiological study that has taken this approach to analyzing the responses of gustatory cells has shown that the across-neuron patterns reflect the qualitative similarities among taste stimuli (Erickson, 1968, 1974, 1982, 1984, 1985; Erickson et al., 1965; Frank, 1991; Frank et al., 1988; Giza & Scott, 1991; Pfaffmann et al., 1976; Scott & Giza, 1990; Smith, 1985; Smith et al., 1979, 1983b; Travers & Smith, 1979; Van Buskirk & Smith, 1981). Often the across-neuron correlations serve as input to a multivariate statistical procedure in order to generate a "taste space," which represents the neural similarities and differences among the stimuli. A taste space for 18 stimuli is shown in Figure 9; this three-dimensional space was derived from the responses of 31 neurons in the hamster PbN and was generated using multidimensional scaling (KYST). Within this space, there is clear separation between the sweet-tasting stimuli (O), the sodium salts (X), the nonsodium salts and acids (●), and the two bitter-tasting stimuli (△). This arrangement of stimuli based on similarities among their across-neuron patterns suggests that there is sufficient information within these patterns to discriminate among these four groups of stimuli, even though any one cell in the hamster PbN is likely to respond to stimuli of more than one group (Smith et al., 1983a, 1983b; Van Buskirk & Smith, 1981).

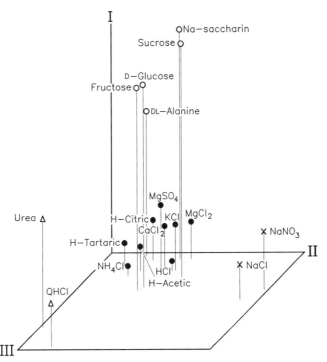

FIGURE 9 Three-dimensional "taste space" showing the similarities and differences among 18 stimuli delivered to the anterior tongue of the hamster. This space was derived from multidimensional scaling (KYST, Bell Laboratories) of the across-neuron correlations among these stimuli recorded from neurons in the parabrachial nuclei of the hamster. Four groups of stimuli are indicated: ○, sweeteners; ×, sodium salts; ●, nonsodium salts and acids; and △, bitter-tasting stimuli. Modified from a figure in Smith, Van Buskirk, Travers, and Bieber (1983b), with permission.

2. Labeled Lines

Although mammalian taste neurons are broadly tuned, many investigators have attempted to group them into functionally meaningful categories (Boudreau & Alev, 1973; Boudreau et al., 1985; Boudreau, Do, Sivakumar, Oravec, & Rodriquez, 1987; Frank, 1973; Frank et al., 1988; Hanamori et al., 1988; Pfaffmann, 1941, 1955; Smith et al., 1983a; Travers & Smith, 1979; Van Buskirk & Smith, 1981). It is obvious from knowledge about the organization of sensitivities in hamster CT and IXth nerves that it is possible to group these taste fibers into classes based on their best stimulus, if four stimuli representing different taste qualities are used. Indeed, a hierarchical cluster analysis of the similarities and differences in the shapes of their response profiles across a broader array of stimuli results in distinct groups of CT fibers (Figure 3) that correspond to these "best-stimulus" groups. The implication of distinct fiber types in the coding of taste quality began with the categorization of hamster CT fibers into best-stimulus groups (Frank, 1973). This categorization became the focus of an ensuing controversy over the neural representation of taste quality when it was proposed that these fiber types code taste quality in a labeled-line fashion (Pfaffmann, 1974; Pfaffmann et al., 1976). This hypothesis suggests that "sweetness" is coded by activity in sucrose-best neurons; "saltiness," by activity in NaCl-best neurons; etc. Thus, in contrast to a "population" approach to taste coding (across-neuron patterns), this labeled-line position advocates a "feature extraction" approach, in which particular neurons (or groups of neurons) play specific roles in the representation of taste quality. Pfaffmann (1974) first proposed a labeled-line code for sweetness because activity in sucrose-best fibers in the squirrel monkey CT nerve correlated better with the animal's preference behavior toward a number of sugars than did activity in the whole nerve. Recently, on the basis of the specific effects of amiloride on the responses to sodium and lithium salts in rat and hamster NaCl-best CT fibers (Hettinger & Frank, 1990; Ninomiya & Funakoshi, 1988) and in rat NaCl-best NST neurons (Giza & Scott, 1991; Scott & Giza, 1990), and on what appear to be the specific effects of sodium deprivation on NaCl-best fibers in the rat CT (Contreras & Frank, 1979), Scott and colleagues (Giza & Scott, 1991; Scott & Giza, 1990; Scott & Plata-Salaman, 1991) have argued that saltiness is coded by activity in NaCl-best neurons. Data showing that amiloride treatment of the tongue prevents rats from discriminating sodium from nonsodium salts (Hill et al., 1990) and produces across-neuron patterns within the NST that cannot distinguish between sodium salts and nonsodium salts or acids (Giza & Scott, 1991; Scott & Giza, 1990) have been interpreted as supporting this argument. On the other hand, human psychophysical studies show that amiloride treatment does not reduce the saltiness of Na^+ or Li^+ salts (Ossebaard & Smith, 1995, 1996). These inconsistencies emphasize the inherent hazards of drawing inferences about taste quality coding from neurophysiological studies in animals.

B. Gustatory Neuron Types: Is Taste Coded by Labeled Lines?

The labeled-line hypothesis requires the existence of neuron types for the coding of taste quality, whereas the across-neuron pattern theory does not. The number of labeled lines would equal the number of discrete taste qualities, which would each be signaled by activity in separate afferent channels. Consequently, the existence of gustatory neuron types has been sharply contested, based on the assumption that their existence somehow implicates them as labeled lines. However, the mere existence of fiber types (defined by their best stimulus, similarities in their profiles, or by other criteria such as their amiloride sensitivity) does not necessarily imply that these classes of cells comprise labeled lines. A classic example where receptor types are evident but where there is general agreement about the existence of a pattern code is in vertebrate color vision (Crick, 1979; Erickson, 1984; Smith, 1985; Smith & Frank, 1993) (as discussed later).

1. Controversy over Fiber Types

Following the introduction of the best-stimulus classification of CT fibers (Frank, 1973), Woolston and Erickson (1979) correctly noted that the subdivision of neurons into groups based on their response to one or a few stimuli could be an arbitrary division of a continuous population of cells. Further, the choice of stimuli and their concentrations by the experimenter could greatly influence the resulting classification. These considerations are particularly applicable to a sensory system such as taste, in which no stimulus continuum has been identified. Because of such considerations, these investigators argued for an approach to the classification of gustatory neurons based on traditional taxonomic procedures (Rowe & Stone, 1977; Tyner, 1975). Using this approach, Erickson and colleagues examined the classification of neurons in the rat CT nerve (Erickson, Covey, and Doetsch, 1980) and NST (Woolston & Erickson, 1979) using hierarchical cluster analysis. The solutions obtained in each case provided no support for distinct neuron types, suggesting that the subdivision of these cells into best-stimulus categories might be an arbitrary exercise. However, the rat CT primarily possesses only two sets of sensitivities: those to sodium salts and those to acids and other electrolytes, which overlap considerably in NaCl- and HCl-best fibers (Frank et al., 1983). These two systems, however, are readily separated pharmacologically with amiloride at both the CT (Hettinger & Frank, 1990; Ninomiya & Funakoshi, 1988) and NST (Giza & Scott, 1991; Scott & Giza, 1990) levels.

As has been discussed, fiber types are readily distinguishable within the CT and IXth nerves of the hamster, based on the relative similarities and differences among their response profiles (Frank et al., 1988; Hanamori et al., 1988). Further, application of multivariate statistical techniques to neurons in the hamster NST and PbN strongly suggests similar neuron types (McPheeters et al., 1990; Smith, 1985; Smith et al., 1983a). Although the recognition of neuron types depends on the strictness of one's criteria when examining a cluster dendrogram (see Scott & Plata-Salaman, 1991), the dendrograms for the hamster CT (e.g., Figure 3) and IXth nerves (Hanamori et al., 1988) strongly suggest fiber types. These hierarchical arrangements are strikingly different from those for SLN fibers (Dickman & Smith, 1988; Smith & Hanamori, 1991), which suggest a continuous distribution of fiber profiles. Classifying the fibers in the CT and IXth nerves as members of distinct fiber types is a decision based on the relative similarities of the profiles within a cluster and the striking differences among the profiles across clusters. At all levels of the hamster gustatory system that have been examined, except in the SLN, there is strong evidence that taste sensitivities are organized into relatively distinct sets of fibers and neurons. Of course, the real issue with respect to sensory coding is what role these fiber types play in the neural code for taste quality. The mere existence of a fiber-type organization of sensitivities is not enough to conclude that they function as labeled lines in the coding of quality.

2. Amiloride Sensitivity—NaCl-Best Cells

As previously noted, one receptor mechanism for sodium salts is an amiloride-sensitive ion channel on the apical membrane of taste receptor cells (Avenet & Lindemann, 1988; DeSimone et al., 1984; Heck et al., 1984; Kinnamon, 1988; Simon et al., 1988). Recent studies have shown that afferent input from this amiloride-sensitive channel is distributed to NaCl-best fibers in the rat (Ninomiya & Funakoshi, 1988) and hamster (Hettinger & Frank, 1990) CT nerve. Responses to NaCl in HCl-best fibers, which are often substantial (Frank, 1973; Frank et al., 1983; Frank et al., 1988), are unaffected by amiloride treatment. Studies of cells in the NST show that this segregation of amiloride-sensitive input to NaCl-best cells is maintained in the second-order neurons (Giza & Scott, 1991; Scott & Giza, 1990; Smith, Liu, & Vogt, 1996), even though convergence occurs between peripheral fibers and NST cells (Sweazey & Smith, 1987; Travers & Norgren, 1991). These results are strong evidence that at least the sodium-sensitive neuron types described in the CT and NST are "natural" types (Rodieck & Brenning, 1983), rather than merely arbitrary classes.

The selective distribution of amiloride-blockable sensitivities into NaCl-best cells has been interpreted as evidence for the labeled-line coding of saltiness (Giza & Scott, 1991; Scott & Giza, 1990; Scott & Plata-Salaman, 1991). These investigators showed that if sodium channels were blocked by amiloride the response to sodium salts was specifically decreased in NaCl-best cells in the rat NST and that a multidimensional taste space generated from the across-neuron correlations following amiloride treatment indicated that NaCl and LiCl were no longer discriminable from nonsodium salts and acids. These neurophysiological findings correspond to behavioral studies, which show that treatment of the tongue with amiloride during conditioning to avoid NaCl interferes with the ability of rats to make subsequent discriminations between sodium and nonsodium salts (Hill et al., 1990). The conclusion drawn from these experiments was that the distinct taste of NaCl is coded by activity in NaCl-best neurons, i.e., by a labeled line or an "independent coding channel" (Scott & Plata-Salaman, 1991). The selective distribution of amiloride-sensitive input into NaCl-best cells is supportive evidence for the concept of neuron types in taste and it suggests a critical role for these cells in coding the taste of sodium salts. However, these data alone do not provide definitive evidence that these neuron types function as labeled lines (see later).

3. The Role of Neuron Types in Defining the Across-Neuron Patterns

The suggestion that NaCl-best cells in the rat NST comprise a labeled line for saltiness is based on the fact that blocking the response of NaCl-best cells with amiloride prevents the neural discrimination between sodium and nonsodium salts, that is, their across-neuron patterns are no longer distinct. Several years ago, we examined the roles played by neuron types in the hamster brain stem in the definitions of the across-neuron patterns (Smith, 1985; Smith et al., 1983a, 1983b). Our conclusions were that the discrimination among stimuli with different tastes (such as sodium and nonsodium salts) depended on comparisons of the activity in different neuron types (such as NaCl- and HCl-best cells) and that one neuron type alone was insufficient to discriminate between stimuli with different taste qualities. This coding mechanism is similar to the coding of stimulus wavelength by the vertebrate visual system, where three types of broadly sensitive photoreceptor pigments are involved (Marks, Dobelle, & MacNichol, 1964). The color of the wavelength of light falling on the retina can be accurately encoded by considering the relative activity in these three photoreceptors, that is, by a pattern (Boynton, 1966, 1971; Erickson, 1968). Deficiencies in one or more of the photoreceptor pigments result in various forms of visual chromatic deficiency, or "color blindness" (Boynton, 1966, 1971). For example, the lack of a long-wavelength photoreceptor results in a confusion in color discrimination (protanopia), where individuals perceive long-wavelength stimuli as one color (yellow) and short wavelength stimuli as another (blue), with a neutral (gray) point at 494 nm (Boynton, 1966, 1971; Graham, Sperling, Hsia, & Coulson, 1961). Data on color-blind individuals show that the absence of any one of the three photoreceptor types results in the inability to discriminate among particular sets of wavelengths (see Smith, 1985; Smith & Frank, 1993, for additional analysis and discussion of this point).

There are no known analogies to color blindness in taste, but the experiments of Scott and colleagues (Giza & Scott, 1991; Scott & Giza, 1990) on the rat NST provide an experimental demonstration of an analogous phenomenon. Blocking the "sodium receptor" (i.e., the NaCl-best neurons) with amiloride results in the inability of the remaining cells to discriminate between sodium and nonsodium salts, that is, their across-neuron patterns are not distinct without the input from the NaCl-best cells. In our previous work on cells in the hamster PbN, we showed that the across-neuron patterns for sodium salts were distinct from those for nonsodium salts and acids only if the activity of both NaCl-best and HCl-best cells was considered (Smith, 1985; Smith et al., 1983b). Without input from NaCl-best neurons (or the HCl-best neurons) there was a lack of separation within the multidimensional taste space between the sodium and the nonsodium salts and acids exactly like that reported after blocking NaCl-best NST cells with amiloride in the rat. That is, the across-fiber patterns for sodium and nonsodium salts were not different from one another; without considering the differential response of two neuron types (NaCl- and HCl-best) to these two classes of stimuli, they were coexistent with the taste space (Smith, 1985; Smith & Frank, 1993; Smith et al., 1983b). All three neuron types (S, N, H) were necessary for the PbN to sort out three groups of stimuli (sweeteners, sodium salts, and nonsodium salts and acids) based on their across-neuron patterns. We interpreted these results to imply that taste quality discrimination depends on a comparison of activity across broadly tuned neuron types, comparable to the coding of color vision by broadly tuned photoreceptors (Smith, 1985; Smith & Frank, 1993; Smith et al., 1983b). We believe that the results of Scott and colleagues following amiloride treatment in the rat are compatible with this interpretation. The conclusion from their work that NaCl-best cells are coding "saltiness" in a labeled-line fashion is analogous to concluding that the long-wavelength photoreceptor is coding "redness" in a labeled-line manner. In both taste and color vision, elimination of any one cell type results in a lack of separation between stimuli of different quality within multidimensional space (Scott & Giza, 1990; Smith, 1985; Smith & Frank, 1993; Smith et al., 1983b) and a loss of behavioral discrimination among the same stimuli (Boynton, 1971; Graham et al., 1961; Hill et al., 1990). Thus we suggest that the coding of taste quality is analogous to color coding, that is, that the code lies in the relative patten of activity across several neuron types.

The results of the hierarchical cluster analysis of hamster CT and IXth nerve fibers, along with the segregation of amiloride-sensitive receptor input into NaCl-best fibers, suggest strongly that there are functional neuron classes in the peripheral gustatory system. Our earlier work on hamster brain stem cells (Smith et al., 1983a, 1983b; Travers & Smith, 1979; Van Buskirk & Smith, 1981) and the effect of blocking NaCl-best neuron responses in the rat NST with amiloride (Giza & Scott, 1991; Scott & Giza, 1990) demonstrate (1) that these classes are preserved in the brain stem and (2) that the neural patterns of activity generated by taste stimuli are dominated by the responses of particular classes of neurons. No one neuron type alone is capable of providing information that can distinguish the across-neuron patterns evoked by dissimilar-tasting compounds. More than one neuron type must contribute to the pattern in order for the patterns evoked by unlike stimuli to be distinct. Thus, in this sense, the various neuron types (sucrose-, NaCl-, HCl-, or QHCl-best) are critically important for the discrimination of taste quality. The NaCl-best cells or HCl-best cells can define the similarities among sodium salts, but activity in both neuron types is required to distinguish sodium salts from nonsodium salts and acids. Behavioral (Hill et al., 1990) and neural (Scott & Giza, 1990) data in rats support this requirement, but there is no evidence to date that NaCl-best cells are labeled lines signaling "saltiness." On the contrary, we think that NaCl-best cells provide a critical part of a pattern across neuron types that codes saltiness; these cells by themselves cannot distinguish sodium salts from other stimuli (Smith, 1985; Smith & Frank, 1993; Smith et al., 1983b). Taste quality discrimination requires the comparison of activity across several neuron types, some of which (e.g., sucrose-best or QHCl-best fibers) are segregated predominantly into different cranial nerves.

V. INTEGRATIVE MECHANISMS OF TASTE

The relationship of taste quality to gustatory hedonics reflects the role of taste in the control of ingestive behavior. Sucrose and other sweet-tasting (hedonically positive) stimuli drive neurons that ultimately connect to cells controlling the ingestion of nutritive substances. Quinine and other aversive (hedonically negative) stimuli feed into neural systems controlling protective reflexes that prevent the ingestion of toxic substances. Recent studies on the relationship between sweet and bitter stimuli suggest that they activate mutually inhibitory systems within the hindbrain. These parallel systems form part of the neural substrate for ingestive and protective systems that can be activated by gustatory stimulation.

A. Special Visceral Afferent System

Whereas taste physiologists have focused largely on the role of gustatory afferent fibers and central neurons in taste quality perception, there are a number of taste-mediated somatic and visceral responses, ranging from tongue movements to salivation to preabsorptive insulin release, which have their neuronal substrate within the brain stem (Grill & Norgren, 1978a, 1978b; Kawamura & Yamamoto, 1978; Mattes, 1987). The differential sensitivities of the VIIth, IXth, and Xth nerves, the organization of their projections into the brain stem, the sensitivities of NST neurons, and the contribution of gustatory afferent input to taste-mediated behaviors are summarized in Figure 10.

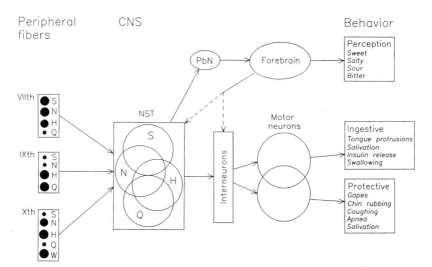

FIGURE 10   Schematic diagram of the chemosensory inputs of three cranial nerves to the taste-responsive portion of the nucleus of the solitary tract (NST), and their putative role in taste-mediated behaviors. The size of the filled circle for each of the peripheral nerves (VIIth, IXth, and Xth) depicts the relative responsiveness of these nerves to sucrose (S), NaCl (N), HCl (H), QHCl (Q), and water (W). Sensitivities of NST cells are largely overlapping, with each cell type somewhat responsive to two or three of the basic stimuli (Travers & Smith, 1979). Sucrose and QHCl stimulate few of the same NST cells, however. Output from the NST ascends in the classic taste pathway via the parabrachial nuclei (PbN) to give rise to perceptions of sweetness, saltiness, sourness, and bitterness and to hedonic tone (not depicted). Local reflex circuits within the brain stem control ingestive and protective responses evoked by taste stimulation. Behavioral data suggest that both ingestive and protective responses can be triggered in parallel, depending on the quality of the stimulus (Brining et al., 1991; Berridge & Grill, 1984). Reprinted from Smith and Frank (1993). Sensory coding by peripheral taste fibers. In S. A. Simon and S. D. Roper (Eds.), *Mechanisms of Taste Transduction* (pp. 295–338). Boca Raton: CRC Press. Reprinted by permission of CRC Press, Boca Raton, Florida.

Taste may be considered the oral component of the visceral afferent system, which includes gustatory, respiratory, cardiovascular, and gastrointestinal functions (Norgren, 1985). Taste buds innervated by the VIIth, IXth, and Xth nerves contribute differentially to this visceral continuum, with the VIIth nerve fibers responsive primarily to hedonically positive stimuli such as sucrose and NaCl, IXth nerve fibers most sensitive to aversive stimuli such as HCl and QHCl, and Xth nerve fibers responsive to stimuli that deviate from the normal pH and ionic milieu of the larynx. Sucrose, for example, stimulates predominantly fibers of the VIIth nerve. These fibers project into the NST, where sucrose-best cells also respond to NaCl and HCl, but are often inhibited by QHCl (Smith & Travers, 1979; Travers & Smith, 1979). Ultimately, the output of these second-order neurons ascends to the forebrain to give rise to the perception of sweetness (Figure 10). Simultaneously, these cells provide input to somatic and visceral motor systems that drive the ingestive components of feeding behavior, including rhythmic mouth movements, tongue protrusions, lateral tongue protrusions, salivary secretion, insulin release, and swallowing (Brining, Belecky, & Smith, 1991; Grill & Norgren, 1978a; Mattes, 1987). Conversely, QHCl stimulates predominantly fibers of the IXth nerve. These fibers project into the NST, where they drive cells that are also responsive to HCl and NaCl but not to sucrose (Giza & Scott, 1991; Sweazey & Smith, 1987). Quinine-sensitive cells of the NST send ascending projections to the forebrain to give rise to sensations of bitterness (Figure 10), but they also provide input to motor systems that drive behaviors such as gaping, chin rubbing, forelimb flailing, and locomotion, associated with rejection (Brining et al., 1991; Grill & Norgren, 1978a). In the rat, the number of gapes elicited by quinine stimulation is reduced by almost one-half after bilateral transection of the IXth nerve (Travers, Grill, & Norgren, 1987).

The superior laryngeal branch of the Xth nerve is involved in swallowing, airway protection, and a number of other visceral reflexes. Respiratory apnea is produced by laryngeal stimulation with water (Boggs & Bartlett, 1982; Storey & Johnson, 1975) and chemosensory fibers of the rat SLN have been shown to mediate diuresis in response to stimulation of the laryngeal mucosa with water (Shingai, Miyaoka, & Shimada, 1988). Besides its obvious role in regulating ingestive behavior, taste also triggers a number of metabolic responses, including salivary, gastric, and pancreatic secretions (Mattes, 1987), although the specific contributions of particular cranial nerves to these responses are not well understood. Thus, besides their mediation of gustatory sensation, taste buds may have a number of roles related to gustatory–visceral regulation, depending on their peripheral distribution and innervation.

Input from taste receptors is critical in the control of ingestive behavior; as noted above, ingestive and protective response sequences can be triggered by gustatory stimulation. Although separating the role of VIIth and IXth nerve fibers in the control of ingestive behavior is difficult in mammals, the differential role of these cranial nerves in controlling ingestive behavior is seen clearly in studies of fish (Atema, 1971; Bardach, Todd, & Crickmer, 1967). The catfish facial system controls the localization and pickup phases of feeding, whereas the vagal system controls the swallowing phase (Atema, 1971). These differential roles are mediated in part by the brain stem connections of the facial and vagal lobes (Finger & Morita, 1985; Morita & Finger, 1985). The situation in mammals is anatomically more complex, involving overlap among the brain stem projections of the VIIth, IXth, and Xth nerves. Nevertheless, it is likely that these separate cranial nerves provide differential control over various aspects of ingestive behavior, even in mammals (Nowlis, 1977).

B. Taste Reactivity

Besides evoking the perception of taste quality, gustatory stimuli trigger several reflexive response sequences that are related to the ingestion and rejection of food substances. These behavioral sequences range from ingestive and protective somatic motor responses to visceral motor activity associated with gastrointestinal function. A number of overt motor responses to taste stimulation can be quantified and measured; these responses have been termed *taste reactivity* and are useful indexes of the hedonic value of a gustatory stimulus (Grill & Norgren, 1978a; Schwartz & Grill, 1984). Taste reactivity responses appear to be mediated by neural substrates within the hindbrain (Grill & Norgren, 1978b).

1. Appetitive and Aversive Responses

Behavioral response sequences triggered by taste stimulation were first quantified in the rat by measures of taste reactivity (Grill & Norgren, 1978a). Taste reactivity in the rat consists of sequences of ingestive or protective behaviors that reflect the palatability (i.e., hedonic value) of a gustatory stimulus. Ingestive responses begin with rhythmic mouth movements, followed by midline tongue protrusions and lateral tongue protrusions; this sequence is triggered by sucrose and other hedonically positive (appetitive) stimuli. Protective responses include oral gapes and a number of somatic motor sequences such as head shaking, chin rubbing, forelimb flailing, paw pushing, and face washing (Grill & Norgren, 1978a), all elicited by quinine and other hedonically negative (aversive) stimuli. In later studies, locomotion, rearing, and active and passive fluid rejection were included in the array of aversive response components (Schwartz & Grill, 1984). Results from taste reactivity tests typically correspond with other short-term palatability measures such as lick-rate tests (Schwartz & Grill, 1984).

In the rat, sucrose and quinine produce opposite patterns of ingestive and protective taste reactivity (Grill & Norgren, 1978a) and a combination of these behaviors can be triggered by mixtures of sucrose and quinine (Berridge & Grill, 1984). Similarly, in the hamster the patterns of taste reactivity to sucrose and quinine are quite different, whereas sodium salts and acids produce patterns consisting of combinations of both ingestive and protective behaviors (Brining et al., 1991). The taste quality of the stimulus is very directly related to the specific pattern of taste reactiv-

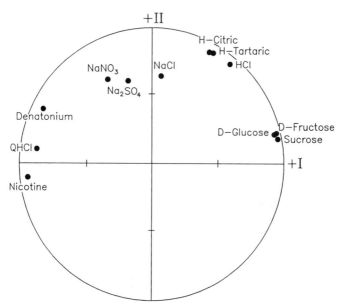

FIGURE 11   Two-dimensional representation of the correlations of hamster taste reactivity profiles with two common factors (I and II). Each point in the space represents the correlation of one stimulus profile with two factors. Zero is at the origin, where the axes cross, and ±0.50 is marked on each axis by crossing line segments. A value of ±1.0 is indicated by the intersections of the circle with the axes. A point would fall on the circumference of the circle if 100% of the variance of a profile were accounted for by these two factors. Stimuli with similar taste quality (e.g., the sugars or the bitter-tasting stimuli) produce similar patterns of taste reactivity, as indicated by their proximity within this factor space. Reprinted from *Physiology & Behavior 49*; Brining, Belecky, and Smith; Taste reactivity in the hamster, 1265–1272. Copyright 1991, with kind permission from Elsevier Science Ltd, The Boulevard, Langford Lane, Kidlington OX5 1GB, UK.

ity that is elicited in the hamster. A two-dimensional representation of the similarities and differences in these patterns for 12 stimuli is shown in Figure 11. This figure shows the correlations of the taste reactivity profile for each stimulus with two common factors, derived from a principal axis factor analysis of the correlations among the profiles (Brining et al., 1991). A point would fall on the circumference of the drawn circle if 100% of the variance of the taste reactivity profile were accounted for by these two factors. The points for the sugars (sucrose, D-fructose, D-glucose) are distributed near the positive end of Factor I, close to the circle. Factor I represents a taste reactivity profile for the highly palatable sugar stimuli. The points for the three bitter-tasting stimuli (denatonium benzoate, nicotine sulfate, and QHCl) are at the opposite end of Factor I, also close to the circle. The bitter-tasting stimuli produce patterns of taste reactivity that are the opposite of the profile represented by Factor I. The profiles for the acids and salts were intermediate on Factor I, consisting of combinations of ingestive and aversive response sequences (Brining et al., 1991). For each group of stimuli, those within a group produced highly correlated patterns of taste reactivity, as evidenced by their proximity within this factor space. These results suggest that the profiles of taste reactivity are a reflection of both the palatability of the stimuli and of their particular taste quality. In other words, stimuli of similar quality produced taste reactivity responses that were similar.

2. Brain Stem Organization of Gustatory Inputs

Decerebrate rats are able to exhibit normal patterns of taste reactivity to stimuli infused into their mouths (Grill & Norgren, 1978b). Both ingestive and aversive oral and somatic motor responses are elicited by gustatory stimulation in these animals, suggesting that the neural substrate important for discriminating among taste stimuli may be intact within the hindbrain. Similarly, anencephalic human neonates produce normal facial expressions in response to gustatory stimulation (Steiner, 1977). Thus information arising from the peripheral gustatory apparatus must project into the brain stem and make appropriate connections at that level to produce the appropriate response sequences to gustatory stimulation. Data presented above show that VIIth and IXth nerve fibers are differentially sensitive, particularly to sweet and bitter stimuli. This differential sensitivity is compatible with the idea that afferent input from the VIIth nerve may be more important for ingestive behavior and that from the IXth nerve for rejection (Nowlis, 1977). However, transecting either the CT or IXth nerves produces only partial effects on ingestive or aversive responses (Travers et al., 1987; Yamamoto, Matsuo, Kiyomitsu, & Kitamura, 1988). Nevertheless, to produce ingestive responses, peripheral fibers carrying information about sweet stimuli, which are numerous in the CT and GSP nerves, must ultimately provide input to a specific set of motor neurons that control tongue protrusions, lateral tongue movements, and swallowing (see Figure 10). Similarly, the oral and somatic motor outputs characteristic of the responses to quinine must arise from motor neurons that receive input arriving at the NST from peripheral taste fibers carrying information about bitter stimuli, which are most numerous in the IXth nerve (Fig. 10). Second-order gustatory cells in the NST project indirectly, through interneurons in the reticular formation, to brainstem motor nuclei that control oromotor responses (Travers, 1985; Travers & Norgren, 1983).

One can view ingestive and protective motor sequences as parallel processes that can be driven one at a time, as by sucrose or quinine, or simultaneously, as by sucrose/quinine mixtures (Berridge & Grill, 1984) or by salts and acids (Figure 11) (Brining et al., 1991). Because most peripheral gustatory fibers and brain stem neurons have multiple sensitivity to different taste qualities, many of the cells contributing to each of these output systems may respond to several stimuli; thus a stimulus such as NaCl or HCl might evoke output in both systems. This multiple sensitivity, however, does not extend to sucrose and quinine, which do not stimulate the same cells (Frank et al., 1988; Hanamori et al., 1988) and which are often mutually inhibitory in brain stem cells (Travers & Smith, 1979; Van Buskirk & Smith, 1981). Thus sweet and bitter stimuli produce relatively specific afferent input and are characterized by specific and opposite patterns of motor output. These motor patterns can be generated by other stimuli to varying degrees and can be modified in the intact animal by conditioning, presumably via descending influences from the forebrain (Grill, 1985; Grill & Norgren, 1978c).

C. Interaction between Appetitive and Aversive Taste Stimuli

The responses to mixtures of stimuli of different taste quality often involve mixture suppression, in which the magnitude of the sensation of either stimulus is less than when the stimulus is presented alone. Recent evidence shows that brain stem cells of the hamster exhibit profound suppression in the response to sucrose when it is mixed with aversive stimuli, such as citric acid or QHCl (Vogt & Smith, 1993a, 1993b). This suppression is sometimes mutual, as in the suppression of the responses to QHCl by mixtures with sucrose. Although there is some peripheral suppression of the response to sucrose by QHCl (Formaker & Frank, 1996) or by HCl or NH_4Cl (Hyman & Frank, 1980) evident in recordings from the CT nerve, there is no evidence of peripheral suppression of the QHCl response by sucrose. Some of the mutual suppression between sucrose and QHCl may reflect inhibitory mechanisms within the brain stem, which are seen in other relationships, such as between VIIth and IXth nerve inputs to the NST.

1. Inhibition of Sucrose Responses

As delineated in the preceding discussion, sucrose and quinine produce opposite patterns of ingestive and protective taste reactivity (Brining et al., 1991). A combination of these behaviors can be triggered by mixtures of sucrose and quinine (Berridge & Grill, 1984), suggesting that these stimuli and their behavioral consequences are processed in parallel. Although mixtures of sucrose and quinine can elicit both behaviors, these behavioral sequences alternate; animals cannot swallow and gape simultaneously (Breslin, Grill, & Spector, 1992). The regulation of such incompatible responses is likely to involve inhibitory circuits, because inhibition is an essential part of antagonistic motor coordination. Electrophysiological evidence shows that in both the NST and PbN sucrose-best cells are often inhibited by quinine stimulation and vice versa (Travers & Smith, 1979; Van Buskirk & Smith, 1981). In cells of the hamster PbN, mixtures of sucrose with either quinine or citric acid, both of which are aversive to hamsters (see Figure 11) (see Brining et al., 1991), result in significantly smaller responses than those evoked by sucrose alone (Smith, Liu, & Vogt, 1994; Vogt & Smith, 1993a, 1993b).

Figure 12 shows a surface plot of the response frequencies of 22 hamster NST cells over time to 0.1 M sucrose, 0.1 M QHCl, and their mixture (Smith et al., 1994). The across-neuron pattern can be seen by reading the surface plot from back to front and the temporal pattern is shown from left to right. Cells are ranked from back to front in order of their response to 0.1 M sucrose (Figure 12A); the same

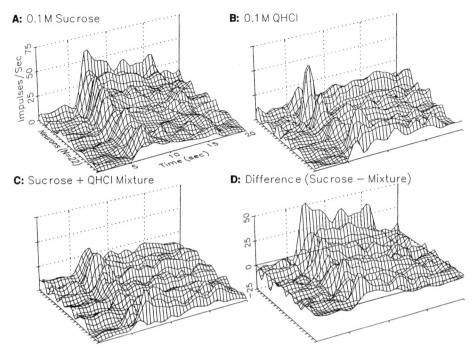

A: 0.1 M Sucrose

B: 0.1 M QHCl

C: Sucrose + QHCl Mixture

D: Difference (Sucrose − Mixture)

FIGURE 12   Responses (impulses/second) of the 22 most sucrose-responsive cells of 29 recorded from the PbN. Surface plots show the response across neurons (back to front) and over time (left to right). Distilled water flowed over the tongue during the time periods 1–5 sec and 16–20 sec; the stimulus occurred during the period 6–15 sec. (A) Responses to 0.1 *M* sucrose; neurons are arranged from back to front in decreasing order of their mean sucrose response (impulses/second over 5 sec). The order of neurons is the same in all plots. (B) Responses of the same cells to 0.1 *M* QHCl. (C) Responses to the mixture of 0.1 *M* sucrose + 0.1 *M* QHCl. (D) Difference in the responses to sucrose and the responses to the mixture (sucrose response minus mixture response). The ordinate in D represents the amount of sucrose suppression. Reprinted from *Physiology & Behavior 56*; Smith, Liu, and Vogt; Neural coding of aversive and appetitive gustatory stimuli: Interactions in the hamster brainstem, 1189–1196. Copyright 1994, with kind permission from Elsevier Science Ltd, The Boulevard, Langford Lane, Kidlington OX5 1GB, UK.

ranking applies throughout the figure. These cells are generally more responsive to sucrose (Figure 12A) than to QHCl (Figure 12B); their response to the mixture (Figure 12C) is considerably smaller than their response to sucrose alone. The magnitude of this suppression is depicted in Figure 12D, where the difference between the sucrose response and the mixture response is shown. An absence of suppression would be a flat surface at the zero value on the ordinate; many of these cells respond as much as 25–50 impulses/sec less to the mixture than to sucrose alone. Across 29 cells the average response to this mixture was significantly less than the response to sucrose. The magnitude of suppression of the sucrose response was positively

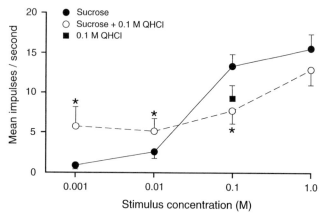

FIGURE 13   Mean responses (±1 SEM) of 29 PbN cells to four concentrations of sucrose alone (●) and mixed with 0.1 *M* QHCl (○). The mean response to 0.1 *M* QHCl is also shown (■). Statistically significant differences (two-tailed *t*-test, *p* < .05) between the mixture response and the response to the more effective component are indicated by asterisks. Reprinted from *Physiology & Behavior 56*; Smith, Liu, and Vogt; Neural coding of aversive and appetitive gustatory stimuli: Interactions in the hamster brainstem, 1189–1196. Copyright 1994, with kind permission from Elsevier Science Ltd, The Boulevard, Langford Lane, Kidlington OX5 1GB, UK.

correlated with the magnitude of the sucrose response, but was independent of the magnitude of the response to QHCl (Smith et al., 1994; Vogt & Smith, 1993a).

Mixing 0.1 *M* QHCl with sucrose resulted in suppression of sucrose responses by QHCl (as in Figure 12), but suppression of the response to QHCl by the weaker sucrose concentrations was also evident (Smith et al., 1994; Vogt & Smith, 1993a), as shown in Figure 13. Here the mean response to each of four concentrations of sucrose (filled circles) is compared to the mean response to the mixture of each of these concentrations with 0.1 *M* QHCl (open circles). The mean response to 0.1 *M* QHCl alone is shown by the filled square. Significant differences from the more effective component are indicated by asterisks (*p* < .05). The responses to the stronger concentrations of sucrose (0.1 and 1.0 *M*) are reduced by the addition of QHCl, although for 1.0 *M* sucrose the difference is not significant. The suppression by QHCl that is evident at 0.1 *M* sucrose is overcome somewhat by 1.0 *M* sucrose. At the lower sucrose concentrations (0.001 and 0.01 *M*), the response to the mixture is significantly less than the response to 0.1 *M* QHCl alone, indicating that sucrose also suppresses the response to QHCl.

2. Inhibitory Interactions in the Brain Stem

The preceding results demonstrate that sucrose and QHCl are mutually suppressive in their effects on taste-responsive cells in the PbN when they are presented as a mixture. The degree to which QHCl suppresses the response to sucrose is correlated with the responsiveness of a cell to sucrose and is independent of the magnitude of its response to QHCl (Vogt & Smith, 1993a); the same is also true of the suppression of PbN sucrose responses by citric acid (Vogt & Smith, 1993b). This implies that the processes activated by QHCl or citric acid that result in sucrose suppression are external to the sucrose-responsive neuron, perhaps arising from inhibitory influences from other cells. Few chorda tympani fibers or central neurons in the hamster are very responsive to anterior tongue stimulation with QHCl (Frank, 1973; Frank et al., 1988; Travers & Smith, 1979; Van Buskirk & Smith, 1981). Nevertheless, these stimuli produce mutual suppression when presented to the anterior tongue as a mixture. Such effects may require convergent inhibitory input onto brain stem neurons.

Evidence from other studies also suggests inhibitory interactions in the brain stem gustatory system. For example, when the influence of the CT nerve is blocked by anesthesia, the magnitude of NST responses to stimulation of the posterior tongue increases significantly (Halpern & Nelson, 1965). In addition, anterior tongue stimuli have been observed to inhibit the resting activity of PbN cells responsive to posterior tongue stimulation (Norgren & Pfaffmann, 1975). Stimulation of the anterior tongue and posterior oral cavity together often produces responses in NST neurons smaller than those to the more effective receptive field stimulated alone (Sweazey & Smith, 1987). Together, these data imply inhibitory interactions between VIIth and IXth nerve inputs. Because the VIIth and IXth nerves are differentially sensitive to sweet and bitter stimuli (Frank, 1973; Frank et al., 1988; Hanamori et al., 1988; Harada & Smith, 1992; Smith & Frank, 1993), these data imply further that parallel appetitive and aversive inputs are mutually inhibitory.

Responses to gustatory stimulation recorded from brain stem cells are subject to several modulatory influences. For example, systemic administration of glucose, insulin, and pancreatic glucagon all produce inhibition of the responses of cells in the rat NST to tongue stimulation with glucose (Giza et al., 1992). The mechanisms underlying these inhibitory effects are unknown; they could involve a direct effect of the increased availability of glucose on the recorded cells or an inhibitory synaptic influence descending from the forebrain. That descending pathways can exert a modulatory influence over brain stem taste cells has been shown by electrophysiological studies on decerebrate rats (Hayama, Ito, & Ogawa, 1985; Mark, Scott, Chang, & Grill, 1988). Reversible blockade of the gustatory neocortex (GN) with procaine results in both increased and decreased responses to taste stimulation in rat PbN neurons (DiLorenzo, 1990), showing that PbN cells are both facilitated and inhibited by descending corticofugal inputs. Electrical stimulation of GN produces both excitatory (via direct synaptic input) and inhibitory influences on PbN cells (DiLorenzo & Monroe, 1992). There are direct descending projections from insular cortex to both PbN (Saper, 1982; Shipley & Sanders, 1982) and NST (Shipley, 1982). We have shown that ipsilateral GN stimulation both excites and inhibits the activity of cells of the rostral NST and that inhibitory responses often can be blocked by the γ-aminobutyric acid A ($GABA_A$) receptor antagonist bicuculline (Liu, Behbehani, & Smith, 1993a).

D. Inhibitory Mechanisms in the Brain Stem

Studies show that the inhibitory neurotransmitter GABA produces inhibition of cells within the gustatory portion of the NST of both rats (Wang & Bradley, 1993) and hamsters (Liu et al., 1993b) in *in vitro* slice preparations of the brain stem. That such an inhibitory mechanism can affect taste-evoked responses has been shown in studies *in vivo* in which GABA produces a differential inhibition of the responses to sucrose in NST cells (Smith et al., 1994). These data suggest that a tonic GABAergic network within the NST could be involved in regulating the relationship between appetitive and aversive stimuli in the control of ingestive and protective response sequences.

1. GABA Inhibition *in Vitro*

Studies have begun to investigate inhibitory mechanisms within the gustatory region of the NST. The inhibitory neurotransmitter GABA has been shown to play a role in the processing of respiratory, cardiovascular, and other information in the visceral portion of the NST (Bennett, McWilliam, & Shepheard, 1987; Bousquet, Feldman, Bloch, & Schwartz, 1982; Jordan, Mifflin, & Spyer, 1988; Wang & Bieger, 1991). Immunocytochemical studies have shown that within the gustatory NST of both rats (Lasiter & Kachele, 1988) and hamsters (Davis, 1993) there are many small ovoid interneurons that express GABA or its degradative enzymes. Similar GABAergic cells are also distributed within caudal regions of the NST (Blessing, Oertel, & Willoughby, 1984; Maley & Newton, 1985). Electrophysiological recordings from cells in the rostral NST in *in vitro* slices from both rats (Wang & Bradley, 1993) and hamsters (Liu et al., 1993b) have shown that GABA produces inhibition of activity in these cells, which is mediated predominantly by the $GABA_A$ receptor subtype.

Extracellular recordings from 80 cells in the rostral central (RC) and rostral lateral (RL) subdivisions of the NST of the hamster *in vitro* show that over two-thirds of the neurons were inhibited by microinjection of GABA near the recorded cell (Liu et al., 1993b). The responses to GABA were dose-dependent and were blocked by the $GABA_A$ receptor antagonist, bicuculline methiodide (BICM). Action potentials recorded from a cell during GABA application are shown in Figure 14A; GABA inhibited the ongoing spontaneous discharge. The rate histogram in Figure 14B shows the dose-dependent responses of a cell, located in the RC subdivision of the NST, to increasing concentrations of GABA, which more completely blocked the spontaneous activity of this cell for a longer duration at the higher GABA concentration. The inhibitory effect of GABA was often eliminated by application of an equivalent dose of BICM (Figure 14C); most inhibitory responses could be blocked by BICM. However, the $GABA_B$ receptor agonist baclofen produced inhibition in a small number of cells, suggesting that some of the inhibition in the rostral NST is $GABA_B$ mediated. Application of BICM alone produced an increase in the baseline

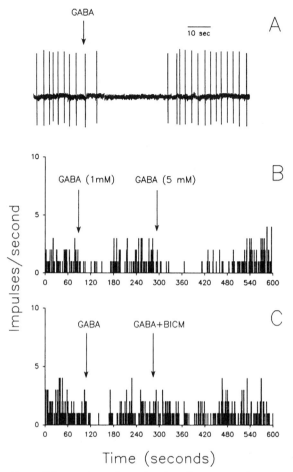

Time (seconds)

FIGURE 14   Extracellular responses of rostral NST cells to γ-aminobutyric acid (GABA) and GABA + bicuculline methiodide (BICM), recorded from an *in vitro* brain stem slice. (A) The inhibitory effect of GABA on the spontaneous impulse frequency of a cell in the rostral central (RC) subdivision of the NST. (B) Extracellular impulse frequency recorded from an NST neuron in the RC subdivision in response to GABA at two different pipette concentrations (1 and 5 m*M*). Bin width = 1 sec. (C) Impulse frequency of the same cell in response to 100 nl of 1 m*M* GABA and to 100 nl of 1 m*M* GABA + 1 m*M* BICM. Reprinted from Liu, Behbehani, and Smith (1993b), *Chemical Senses 18,* by permission of Oxford University Press.

discharge of many NST cells. This increase did not occur in Ca²⁺-free bathing medium, which blocks synaptic release within the slice. These results suggest that the gustatory portion of the NST is under the influence of a tonic GABAergic inhibitory network. Indeed, in patch–clamp experiments, many spontaneous inhibitory postsynaptic currents (IPSCs) were detected (Liu et al., 1993b).

2. Inhibition of Sucrose Responses by GABA

These *in vitro* data (Liu et al., 1993b) and data from rats (Wang & Bradley, 1993) strongly implicate a GABAergic inhibitory mechanism in the processing of information through the gustatory portion of the NST. Such a mechanism could be involved in the inhibitory interactions that have been observed in NST and PbN neurons. However, from *in vitro* studies alone one cannot draw conclusions about the functional significance of such a system, or even that GABA influences taste (as opposed to somatosensory or visceral) responses. Therefore, pharmacological manipulations *in vivo* are necessary to relate these *in vitro* observations to gustatory processing. The action of GABA and its antagonist BICM on taste-elicited responses of cells in the NST has been demonstrated by extracellular recording combined with local micropressure injection of these agents into the nucleus (Smith et al., 1994).

In an initial experiment, the spontaneous activity of 10 of 19 cells was inhibited by GABA in a dose-dependent fashion. BICM blocked the GABA-induced inhibition of 4 of 6 cells tested. Peristimulus time histograms depicting the responses of one NST cell are shown in Figure 15. Each panel in Figure 15 shows 10 min of continuous neural activity. This neuron responded best to NaCl (N), but also showed responses to KCl (K) and sucrose (S), as shown in Figure 15A. This cell had a relatively high rate of spontaneous activity (> 10 impulses/sec), which was inhibited by 1 mM GABA; this inhibition was blocked by 2 mM BICM (Figures 15B and 15C). GABA inhibition persisted for about 2 min after the 10-msec GABA pulse. Stimulation of the anterior tongue with sucrose, NaCl, or KCl following GABA administration resulted in reduced responses, but not if BICM was also microinjected (GABA + BICM; Figure 15C). The responses evoked by taste stimulation were stable over the recording session and recovered completely after drug administration, as shown by the responses to sucrose, KCl, and NaCl at the end of the experiment (Figure 15D).

The activity elicited by gustatory stimuli was significantly inhibited by GABA in 8 of these 19 cells. The response to sucrose was significantly inhibited by GABA in all 8 cells, with an average decrease of 56% in impulse frequency. Inhibition of the KCl response occurred in 5 of these 8 neurons; the mean decrease in firing rate was 39%. GABA produced an inhibition in the response to NaCl in only 3 cells, with an average decrease of 17%. Cells in all three best-stimulus classes (sucrose-, NaCl-, or KCl-best) were inhibited by GABA; the degree of inhibition related to the size of the response to a particular stimulus rather than to the cell's best-stimulus category. For example, the 7 cells most responsive to sucrose were among those 8 in which the sucrose response was inhibited, although there were only 4 sucrose-best cells in this sample of neurons.

The inhibition of taste-evoked responses by GABA was dose dependent. In an additional sample of 17 cells, the response of each cell's best stimulus was recorded before and after administration of GABA at three doses. For this experiment, the

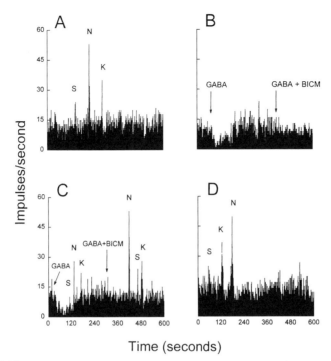

Time (seconds)

FIGURE 15   Peristimulus time histograms showing the *in vivo* responses of an NST cell to 0.1 *M* sucrose, 0.032 *M* NaCl, and 0.1 *M* KCl. In each panel, 10 min of ongoing activity is depicted. (A) Control responses to sucrose (S), NaCl (N), and KCl (K); this cell was NaCl-best. (B) Inhibition of ongoing spontaneous activity by micropressure injection of 1 m*M* GABA (50 nl) and blockage of this inhibition by 2 m*M* BICM (GABA + BICM). (C) Inhibition of responses to S, N, and K stimulation after microinjection of GABA, but not after GABA + BICM. (D) Control responses to S, N, and K at the end of the experiment. Reprinted from *Physiology & Behavior 56*; Smith, Liu, and Vogt; Neural coding of aversive and appetitive gustatory stimuli: Interactions in the hamster brainstem, 1189–1196. Copyright 1994, with kind permission from Elsevier Science Ltd, The Boulevard, Langford Lane, Kidlington OX5 1GB, UK.

effective dose of GABA was controlled by varying the duration of the Picospritzer pulse (5, 10, or 20 msec) that delivered the microwpressure injection. As in the first set of experiments, the response to sucrose was more suppressed by GABA administration than that to KCl or NaCl. However, the responses to all three stimuli were inhibited by GABA in a dose-dependent fashion, showing greater inhibition at longer durations of microinjection. The mean percentage inhibition of the response of each best-stimulus class of cells to its best stimulus after varying durations of GABA microinjection is shown in Figure 16. The amount of inhibition of the sucrose response is about twice that of the responses to NaCl or KCl. As in the first experiment, 10-msec pressure pulse of 1 m*M* GABA produced about a 60% decrement in the response to sucrose, but less than 30% in the responses to

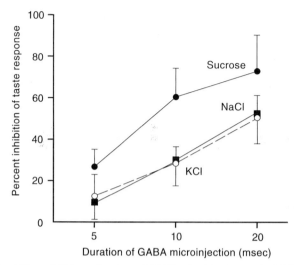

FIGURE 16   Effects of different durations of GABA microinjection on the inhibition of the responses of hamster NST cells to sucrose, NaCl, and KCl. The mean percentage inhibition of each response is shown on the ordinate. Error bars show ±1 SEM. Reprinted from *Physiology & Behavior 56*; Smith, Liu, and Vogt; Neural coding of aversive and appetitive gustatory stimuli: Interactions in the hamster brainstem, 1189–1196. Copyright 1994, with kind permission from Elsevier Science Ltd, The Boulevard, Langford Lane, Kidlington OX5 1GB, UK.

NaCl and KCl. A 20-msec pulse produced almost an 80% decrement in the sucrose response.

These latter data and those recorded *in vitro* strongly implicate a GABAergic inhibitory mechanism in the processing of gustatory information through the NST. The electrophysiological data recorded from brain stem slice preparations in the hamster (Liu et al., 1993b) and rat (Wang & Bradley, 1993) have shown that cells within the gustatory regions of the rostral NST are modulated by what appears to be a tonic GABAergic network. These *in vivo* results show that GABA exerts an inhibitory influence over the responses of gustatory cells; it suppresses the spontaneous activity of many of these cells and produces a decrease in the response to tastants delivered to the anterior tongue. The greater inhibition of the response to sucrose by GABA microinjection into the NST parallels the greater inhibition of sucrose seen in mixtures of sucrose with QHCl (Smith et al., 1994; Vogt & Smith, 1993a) or with citric acid (Vogt & Smith, 1993b). Such a differential effect is also compatible with the differential inhibition of the response to glucose by NST cells after systemic administration of glucose, pancreatic glucagon, or insulin (Giza et al., 1992). The inhibitory effects of aversive stimuli like QHCl and acid on the responses of brain stem cells to sucrose may be mediated by this GABAergic network. Studies directed at the ability of GABA antagonists to block these inhibitory effects can test this hypothesis directly.

References

Altschuler, S. M., Bao, X., Bieger, D., Hopkins, D. A., & Miselis, R. R. (1989). Viscerotopic representation of the upper alimentary tract in the rat: Sensory ganglia and nuclei of the solitary and spinal trigeminal tracts. *Journal of Comparative Neurology, 283,* 248–268.

Atema, J. (1971). Structures and functions of the sense of taste in the catfish (*Ictalurus natalis*). *Brain, Behavior and Evolution, 4,* 273–294.

Avenet, P., & Lindemann, B. (1988). Amiloride-blockable sodium currents in isolated taste receptor cells. *Journal of Membrane Biology, 105,* 245–255.

Bardach, J. E., Todd, J. H., & Crickmer, R. (1967). Orientation by taste in fish of the genus *Ictalurus*. *Science, 155,* 1276–1278.

Bartoshuk, L. M. (1971). The chemical senses. I. Taste. In J. W. King & L. A. Riggs (Eds.), *Woodworth and Schlosberg's experimental psychology* (pp. 169–191). New York: Holt, Rinehart and Winston.

Bartoshuk, L. M., Murphy, C., & Cleveland, C. T. (1978). Sweet taste of dilute NaCl: Psychophysical evidence for a sweet stimulus. *Physiology & Behavior, 21,* 609–613.

Bartoshuk, L. M., Rennert, K., Rodin, J., & Stevens, J. C. (1982). Effects of temperature on the perceived sweetness of sucrose. *Physiology & Behavior, 28,* 905–910.

Beckstead, R., Morse, J., & Norgren, R. (1980). The nucleus of the solitary tract in the monkey: Projections to the thalamus and brain stem nuclei. *Journal of Comparative Neurology, 190,* 259–282.

Beebe-Center, J. G., Black, P., Hoffman, A. C., & Wade, M. (1948). Relative per diem consumption as a measure of preference in the rat. *Journal of Comparative and Physiological Psychology, 41,* 239–251.

Belecky, T. L., & Smith, D. V. (1990). Postnatal development of palatal and laryngeal taste buds in the hamster. *Journal of Comparative Neurology, 293,* 646–654.

Bennett, J. A., McWilliam, P. N., & Shepheard, S. L. (1987). A γ-aminobutyric-acid-mediated inhibition of neurones in the nucleus tractus solitarius of the cat. *Journal of Physiology (London), 392,* 417–430.

Berridge, K. C., & Grill, H. J. (1984). Isohedonic tastes support a two-dimensional hypothesis of palatability. *Appetite, 4,* 221–231.

Bertino, M., Beauchamp, G. K., & Engelman, K. (1986). Increasing dietary salt alters salt taste preference. *Physiology & Behavior, 38,* 203–213.

Bieber, S. L., & Smith, D. V. (1986). Multivariate analysis of sensory data: A comparison of methods. *Chemical Senses, 11,* 19–47.

Blessing, W. W., Oertel, W. H., & Willoughby, J. O. (1984). Glutamic acid decarboxylase immunoreactivity is present in perikarya of neurons in nucleus tractus solitarius of rat. *Brain Research, 322,* 346–350.

Boggs, D. F., & Bartlett, D., Jr. (1982). Chemical specificity of a laryngeal apneic reflex in puppies. *Journal of Applied Physiology, 53,* 455–462.

Boudreau, J. C., & Alev, N. (1973). Classification of chemoresponsive tongue units of the cat geniculate ganglion. *Brain Research, 54,* 157–175.

Boudreau, J. C., Do, L. T., Sivakumar, L., Oravec, J., & Rodriquez, C. A. (1987). Taste systems of the petrosal ganglion of the rat glossopharyngeal nerve. *Chemical Senses, 12,* 437–458.

Boudreau, J. C., Sivakumar, L., Do, L. T., White, T. D. Oravec, J., & Hoang, N. K. (1985). Neurophysiology of geniculate ganglion (facial nerve) taste systems: Species comparisons. *Chemical Senses, 10,* 89–127.

Bousquet, P., Feldman, J., Bloch, R., & Schwartz, J. (1982). Evidence for a neuromodulatory role of GABA at the first synapse of the baroreceptor reflex pathway: Effects of GABA derivatives injected into the NTS. *Naunyn-Schmiedeberg's Archives of Pharmacology, 319,* 168–171.

Boynton, R. M. (1966). Vision. In J. Sidowski (Ed.), *Experimental methods and instrumentation in psychology* (pp. 273–330). New York: McGraw-Hill.

Boynton, R. M. (1971). Color vision. In J. W. Kling & L. A. Riggs (Eds.), *Woodworth and Schlosberg's experimental psychology* (pp. 315–368). New York: Holt, Rinehart, and Winston.

Bradley, R. M. (1972). Development of the taste bud and gustatory papillae in human fetuses. In J. F.

Bosma (Ed.), *Third symposium on oral sensation and perception: The mouth of the infant* (pp. 137–162). Springfield, IL: Thomas.

Bradley, R. M. (1982). The role of epiglottal and lingual chemoreceptors: A comparison. In J. E. Steiner & J. R. Ganchrow (Eds.), *Determination of behavior by chemical stimuli* (pp. 37–45). London: Information Retrieval.

Bradley, R., Cheal, M., & Kim, Y. (1980). Quantitative analysis of developing epiglottal taste buds in sheep. *Journal of Anatomy, 130,* 25–32.

Bradley, R. M., Stedman, H. M., & Mistretta, C. M. (1983). Superior laryngeal nerve response patterns to chemical stimulation of sheep epiglottis. *Brain Research, 276,* 81–93.

Brand, J. G., Teeter, J. H., & Silver, W. L. (1985). Inhibition by amiloride of chorda tympani responses evoked by monovalent salts. *Brain Research, 334,* 207–214.

Breslin, P. A. S., Grill, H. J., & Spector, A. C. (1992). A quantitative comparison of taste reactivity behaviors to sucrose before and after lithium chloride pairings: A unidimensional account of palatability. *Behavioral Neuroscience, 106,* 820–836.

Brining, S. K., Belecky, T. L., & Smith, D. V. (1991). Taste reactivity in the hamster. *Physiology & Behavior, 49,* 1265–1272.

Carpenter, J. A. (1956). Species differences in taste preference. *Journal of Comparative and Physiological Psychology, 49,* 139–144.

Chang, F.-C. T., & Scott, T. R. (1984). Conditioned taste aversions modify neural responses to gustatory stimuli. *Journal of Neuroscience, 4,* 1850–1862.

Cleaton-Jones, P. (1971). A denervation study of taste buds in the soft palate of the albino rat. *Archives of Oral Biology, 21,* 79–82.

Contreras, R. J., Beckstead, R. M., & Norgren, R. (1982). The central projections of the trigeminal, facial, glossopharyngeal and vagus nerves: An autoradiographic study in the rat. *Journal of the Autonomic Nervous System, 6,* 303–322.

Contreras, R., & Frank, M. (1979). Sodium deprivation alters neural responses to gustatory stimuli. *Journal of General Physiology, 73,* 569–594.

Crick, F. H. C. (1979). Thinking about the brain. *Scientific American, 241,* 219–232.

Davis, J. D. (1973). The effectiveness of some sugars in stimulating licking behavior in the rat. *Physiology & Behavior, 11,* 39–45.

Davis, B. J. (1993). GABA-like immunoreactivity in the gustatory zone of the nucleus of the solitary tract in the hamster: Light and electron microscopic studies. *Brain Research Bulletin, 30,* 69–77.

DeSimone, J. A., & Ferrell, F. (1985). Analysis of amiloride inhibition of chorda tympani taste response of rat to NaCl. *American Journal of Physiology, 249,* R52–R61.

DeSimone, J. A., Heck, G. L., Mierson, S., & DeSimone, S. K. (1984). The active ion transport properties of canine lingual epithelia in vitro. Implications for gustatory transduction. *Journal of General Physiology, 83,* 633–656.

Dickman, J. D., & Smith, D. V. (1988). Response properties of fibers in the hamster superior laryngeal nerve. *Brain Research, 450,* 25–38.

Dickman, J. D., & Smith, D. V. (1989). Topographic distribution of taste responsiveness in the hamster medulla. *Chemical Senses, 14,* 231–247.

DiLorenzo, P. M. (1989). Across unit patterns in the neural response to taste: Vector space analysis. *Journal of Neurophysiology, 62,* 823–833.

DiLorenzo, P. M. (1990). Corticofugal influence on taste responses in the parabrachial pons of the rat. *Brain Research, 530,* 73–84.

DiLorenzo, P. M., & Monroe, S. (1992). Corticofugal input to taste-responsive units in the parabrachial pons. *Brain Research Bulletin, 29,* 925–930.

Dzendolet, E., & Meiselman, H. L. (1967). Gustatory quality changes as a function of solution concentration. *Physiology & Behavior, 2,* 29–33.

Erickson, R. P. (1963). Sensory neural patterns and gustation. In Y. Zotterman (Ed.), *Olfaction and taste* (pp. 205–213). Oxford: Pergamon.

Erickson, R. P. (1968). Stimulus coding in topographic and non-topographic afferent modalities: On the significance of the activity of individual sensory neurons. *Psychological Review, 75,* 447–465.

Erickson, R. P. (1970). Parallel "population" neural coding in feature extraction. In F. O. Schmitt & F. G. Worden (Eds.), *The neurosciences: Third study program* (pp. 155–170). Cambridge, MA: MIT Press.

Erickson, R. P. (1977). The role of "primaries" in taste research. In J. LeMagnen & P. MacLeod (Eds.), *Olfaction and taste VI* (pp. 369–376). London: Information Retrieval.

Erickson, R. P. (1982). The "across-fiber pattern" theory: An organizing principle for molar neural function. In W. D. Neff (Ed.), *Contributions to Sensory Physiology: Vol. 6* (pp. 79–110). New York: Academic Press.

Erickson, R. P. (1984). On the neural basis of behavior. *American Scientist, 72,* 233–241.

Erickson, R. P. (1985). Grouping in the chemical senses. *Chemical Senses, 10,* 333–340.

Erickson, R. P., Covey, E., & Doetsch, G. S. (1980). Neuron and stimulus typologies in the rat gustatory system. *Brain Research, 196,* 513–519.

Erickson, R. P., Doetsch, G. S., & Marshall, D. A. (1965). The gustatory neural response function. *Journal of General Physiology, 49,* 247–263.

Elliott, R. (1937). Total distribution of taste buds on the tongue of the kitten at birth. *Journal of Comparative Neurology, 66,* 361–373.

Feindel, W. (1956). The neural pattern of the epiglottis. *Journal of Comparative Neurology, 105,* 269–280.

Finger, T. E., & Morita, Y. (1985). Two gustatory systems: Facial and vagal gustatory nuclei have different brainstem connections. *Science, 227,* 776–778.

Fish, H., Malone, P., & Richter, C. (1944). The anatomy of the tongue of the domestic Norway rat. I. The skin of the tongue; the various papillae; their number and distribution. *Anatomical Record, 89,* 429–440.

Fishman, I. Y. (1957). Single fiber gustatory impulses in rat and hamster. *Journal of Cellular and Comparative Physiology, 49,* 319–334.

Formaker, B. K., & Frank, M. E. (1996). Responses of the hamster chorda tympani nerve to binary component taste stimuli—evidence for peripheral gustatory mixture interactions. *Brain Research, 727,* 79–90.

Formaker, B. K., & Hill, D. L. (1988). An analysis of residual NaCl taste response after amiloride. *American Journal of Physiology, 255,* R1002–R1007.

Formaker, B. K., & Hill, D. L. (1991). Lack of amiloride sensitivity in SHR and WKY glossopharyngeal taste responses to NaCl. *Physiology & Behavior, 50,* 765–769.

Frank, M. (1973). An analysis of hamster afferent taste nerve response functions. *Journal of General Physiology, 61,* 588–618.

Frank, M. E. (1991). Taste-responsive neurons of the glossopharyngeal nerve of the rat. *Journal of Neurophysiology, 65,* 1452–1463.

Frank, M. E., Bieber, S. L., & Smith, D. V. (1988). The organization of taste sensibilities in hamster chorda tympani nerve fibers. *Journal of General Physiology, 91,* 861–896.

Frank, M. E., Contreras, R. J., and Hettinger, T. P. (1983). Nerve fibers sensitivie to ionic taste stimuli in chorda tympani of the rat. *Journal of Neurophysiology, 50,* 941–960.

Frank, M. E., & Nowlis, G. H. (1989). Learned aversions and taste qualities in hamsters. *Chemical Senses, 14,* 379–394.

Ganchrow, J. R., & Erickson, R. P. (1970). Neural correlates of gustatory intensity and quality. *Journal of Neurophysiology, 33,* 768–783.

Gill II, J. M., & Erickson, R. P. (1985). Neural mass differences in gustation. *Chemical Senses, 10,* 531–548.

Giza, B. K., & Scott, T. R. (1983). Blood glucose selectively affects taste-evoked activity in the rat nucleus tractus solitarius. *Physiology & Behavior, 31,* 643–650.

Giza, B. K., & Scott, T. R. (1987). Intravenous insulin infusions in rats decrease gustatory-evoked responses to sugars. *American Journal of Physiology, 252,* R994–R1002.

Giza, B. K., & Scott, T. R. (1991). The effect of amiloride on taste-evoked activity in the nucleus tractus solitarius of the rat. *Brain Research, 550,* 247–256.

Giza, B. K., Scott, T. R., & Vanderweele, D. A. (1992). Administration of satiety factors and gustatory responsiveness in the nucleus tractus solitarius of the rat. *Brain Research Bulletin, 28,* 637–639.

Glenn, J. F., & Erickson, R. P. (1976). Gastric modulation of gustatory afferent activity. *Physiology & Behavior, 16,* 561–568.

Gordon, K. D., & Caprio, J. (1985). Taste responses to amino acids in the southern leopard frog, *Rana sphenocephala. Comparative Biochemistry and Physiology, 81A,* 525–530.

Graham, C. H., Sperling, H. G., Hsia, Y., & Coulson, A. H. (1961). The determination of some visual functions of a unilaterally color-blind subject: Methods and results. *Journal of Psychology, 51,* 3–32.

Green, B. G., & Frankmann, S. P. (1987). The effect of cooling the tongue on the perceived intensity of taste. *Chemical Senses, 12,* 609–619.

Grill, H. J. (1985). Introduction: Physiological mechanisms in conditioned taste aversion. *Annals of the New York Academy of Sciences, 443,* 67–88.

Grill, H. J., & Norgren, R. (1978a). The taste reactivity test. I. Mimetic responses to gustatory stimuli in neurologically normal rats. *Brain Research, 143,* 263–280.

Grill, H. J., & Norgren, R. (1978b). The taste reactivity test. II. Mimetic responses to gustatory stimuli in chronic thalamic and decerebrate rats. *Brain Research, 143,* 281–297.

Grill, H. J., & Norgren, R. (1978c). Chronically decerebrate rats demonstrate satiation but not bait shyness. *Science, 201,* 267–269.

Guth, L. (1957). Effects of glossopharyngeal nerve transection on the circumvallate papilla of the rat. *Anatomical Record, 128,* 715–731.

Halpern, B. P. (1967). Chemotopic coding for sucrose and quinine hydrochloride in the nucleus of the fasciculus solitarius. In T. Hayashi (Ed.), *Olfaction and taste II* (pp. 549–562). Oxford: Pergamon.

Halpern, B. P., & Nelson, L. M. (1965). Bulbar gustatory responses to anterior and to posterior tongue stimulation in the rat. *American Journal of Physiology, 209,* 105–110.

Halsell, C. (1992). Organization of parabrachial nucleus efferents to the thalamus and amygdala in the golden hamster. *Journal of Comparative Neurology, 317,* 57–78.

Hamilton, R. B., & Norgren, R. (1984). Central projections of gustatory nerves in the rat. *Journal of Comparative Neurology, 222,* 560–577.

Hanamori, T., Ishiko, N., & Smith, D. V. (1987). Multimodal responses of taste neurons in the frog nucleus tractus solitarius. *Brain Research Bulletin, 18,* 87–97.

Hanamori, T., Miller, I. J., Jr., & Smith, D. V. (1988). Gustatory responsiveness of fibers in the hamster glossopharyngeal nerve. *Journal of Neurophysiology, 60,* 478–498.

Hanamori, T., & Smith, D. V. (1986). Central projections of the hamster superior laryngeal nerve. *Brain Research Bulletin, 16,* 271–279.

Hanamori, T., & Smith, D. V. (1989). Gustatory innervation in the rabbit: Central distribution of sensory and motor components of the chorda tympani, glossopharyngeal, and superior laryngeal nerves. *Journal of Comparative Neurology, 282,* 1–14.

Harada, S., & Smith, D. V. (1992). Gustatory sensitivities of the hamster's soft palate. *Chemical Senses, 17,* 37–51.

Hayama, T., Ito, S., & Ogawa, H. (1985). Responses of solitary tract nucleus neurons to taste and mechanical stimulation of the oral cavity in decerebrate rats. *Experimental Brain Research, 60,* 235–242.

Heck, G. L., Mierson, S., & DeSimone, J. A. (1984). Salt taste transduction occurs through an amiloride-sensitive sodium transport pathway. *Science, 223,* 403–405.

Hellekant, G., DuBois, G. E., Roberts, T. W., & van der Wel, H. (1988). On the gustatory effects of amiloride in the monkey (*Macaca mulatta*). *Chemical Senses, 13,* 89–93.

Herbert, H., Moga, M. M., & Saper, C. B. (1990). Connections of the parabrachial nucleus with the nucleus of the solitary tract and the medullary reticular formation in the rat. *Journal of Comparative Neurology, 293,* 540–580.

Herness, M. S. (1987). Effects of amiloride on bulk flow and iontophoretic taste simuli in the hamster. *Journal of Comparative Physiology A, 160,* 281–288.

Hettinger, T. P., & Frank, M. E. (1990). Specificity of amiloride inhibition of hamster taste responses. *Brain Research, 513,* 24–34.

Hill, D. L. (1987). Development of amiloride sensitivity in the rat peripheral gustatory system: A single fiber analysis. In S. D. Roper & J. Atema (Eds.), *Olfaction and taste IX* (pp. 369–372). New York: New York Academy of Sciences.

Hill, D. L., Formaker, B. K., & White, K. S. (1990). Perceptual characteristics of the amiloride-suppressed sodium chloride taste response in the rat. *Behavioral Neuroscience, 104,* 734–741.

Hosley, M. A., Hughes, S. E., & Oakley, B. (1987). Neural induction of taste buds. *Journal of Comparative Neurology, 260,* 224–232.

Hyman, A. M., & Frank, M. E. (1980). Sensitivities of single nerve fibers in the hamster chorda tympani nerve to mixtures of taste stimuli. *Journal of General Physiology, 62,* 348–356.

Jordan, D., Mifflin, S. W., & Spyer, K. M. (1988). Hypothalamic inhibition of neurones in the nucleus tractus solitarius of the cat is GABA mediated. *Journal of Physiology (London), 399,* 389–404.

Kanwal, J. S., & Caprio, J. (1983). An electrophysiological investigation of the oro-pharyngeal (IX-X) taste system in the channel catfish, *Ictalurus punctatus. Journal of Comparative Physiology A, 150,* 345–357.

Kawamura, Y., & Yamamoto, T. (1978). Studies on neural mechanisms of the gustatory-salivary reflex in rabbits. *Journal of Physiology (London), 285,* 35–47.

Khaisman, E. B. (1976). Particular features of the innervation of taste buds of the epiglottis in monkeys. *Acta Anatomica, 95,* 101–115.

Kinnamon, S. C. (1988). Taste transduction: A diversity of mechanisms. *Trends in Neuroscience, 11,* 491–496.

Kosar, E., Grill, H. J., & Norgren, R. (1986a). Gustatory cortex in the rat. I. Physiological properties and cytoarchitecture. *Brain Research, 379,* 329–341.

Kosar, E., Grill, H. J., & Norgren, R. (1986b). Gustatory cortex in the rat. II. Thalamocortical projections. *Brain Research, 379,* 342–352.

Krimm, R. F., Nejad, M. S., Smith, J. C., Miller, I. J., Jr., & Beidler, L. M. (1987). The effect of bilateral sectioning of the chorda tympani and the greater superficial petrosal nerves on the sweet taste in the rat. *Physiology & Behavior, 41,* 495–501.

Lasiter, P. S., & Kachele, D. L. (1988). Organization of GABA and GABA-transaminase containing neurons in the gustatory zone of the nucleus of the solitary tract. *Brain Research Bulletin, 21,* 623–636.

Liu, H., Behbehani, M., & Smith, D. V. (1993a). The influence of gustatory cortex on taste cells of the solitary nucleus of the hamster. *Neuroscience Abstracts, 19,* 1430.

Liu, H., Behbehani, M., & Smith, D. V. (1993b). The influence of GABA on cells in the gustatory region of the hamster solitary nucleus. *Chemical Senses, 18,* 285–305.

London, R. M., Snowdon, C. T., & Smithana, J. M. (1979). Early experience with sour and bitter solutions increases subsequent ingestion. *Physiology & Behavior, 22,* 1149–1155.

Maes, F. W. (1985). Improved best-stimulus classification of taste neurons. *Chemical Senses, 10,* 35–44.

Maley, B., & Newton, B. W. (1985). Immunohistochemistry of γ-aminobutyric acid in the cat nucleus tractus solitarius. *Brain Research, 330,* 364–368.

Mark, G. P., Scott, T. R., Chang, F.-C. T., & Grill, H. J. (1988). Taste responses in the nucleus tractus solitarius of the chronic decerebrate rat. *Brain Research, 443,* 137–148.

Marks, W. B., Dobelle, W. H., & MacNichol, E. F., Jr. (1964). Visual pigments of single primate cones. *Science, 143,* 1181–1183.

Mattes, R. D. (1987). Sensory influences on food intake and utilization in humans. *Human Nutrition and Applied Nutrition, 41A,* 77–95.

McBurney, D. H. (1974). Are there primary tastes for man? *Chemical Senses and Flavor, 1,* 17–28.

McBurney, D. H., & Gent, G. F. (1979). On the nature of taste qualities. *Psychological Bulletin, 86,* 151–167.

McPheeters, M., Hettinger, T. P., Nuding, S. C., Savoy, L. D., Whitehead, M. C., & Frank, M. E. (1990). Taste-responsiveness neurons and their locations in the solitary nucleus of the hamster. *Neuroscience, 34,* 745–758.

Miller, I. J., Jr. (1977). Gustatory receptors of the palate. In Y. Katsuki, M. Sato, S. Takagi, and Y. Oomura (Eds.), *Food intake and chemical senses* (pp. 173–186). Tokyo: Univ. of Tokyo Press.

Miller, I. J., Jr., & Bartoshuk, L. M. (1991). Taste perception, taste bud distribution, and spatial relationships. In T. V. Getchell, R. L. Doty, L. M. Bartoshuk, & J. B. Snow, Jr. (Eds.), *Smell and taste in health and disease* (pp. 205–233). New York: Raven.

Miller, I. J., Jr., & Smith, D. V. (1984). Quantitative taste bud distribution in the hamster. *Physiology & Behavior, 32,* 275–285.

Miller, I. J., Jr., & Smith, D. V. (1989). Proliferation of taste buds in the foliate and vallate papillae of postnatal hamsters. *Growth, Development and Aging, 52,* 123–131.

Miller, I. J., Jr., & Spangler, K. M. (1982). Taste bud distribution and innervation on the palate of the rat. *Chemical Senses, 7,* 99–108.

Mistretta, C. M., & Bradley, R. M. (1983). Developmental changes in taste responses from glossopharyngeal nerve in sheep and comparisons with chorda tympani responses. *Developmental Brain Research, 11,* 107–117.

Moore, K. L. (1992). *Clinically oriented anatomy* (3rd ed.). Baltimore, MD: Williams & Wilkins.

Morita, Y., & Finger, T. E. (1985). Reflex connections of the facial and vagal gustatory systems in the brainstem of the bullhead catfish, *Ictalurus nebulosus. Journal of Comparative Neurology, 231,* 547–558.

Morrison, G. R. (1967). Behavioural response patterns to salt stimuli in the rat. *Canadian Journal of Psychology, 21,* 141–152.

Moskowitz, H. R., & Arabie, P. (1970). Taste intensity as a function of stimulus concentration and solvent viscosity. *Journal of Texture Studies, 1,* 502–510.

Nachman, M. (1963). Learned aversion to the taste of lithium chloride and generalization to other salts. *Journal of Comparative and Physiological Psychology, 56,* 343–349.

Nejad, M. S. (1986). The neural activities of the greater superficial petrosal nerve of the rat in response to chemical stimulation of the palate. *Chemical Senses, 11,* 283–293.

Nilsson, B. (1979). The occurrence of taste buds in the palate of human adults as evidenced by light microscopy. *Acta Odontologica Scandinavia, 37,* 253–258.

Ninomiya, Y., & Funakoshi, M. (1988). Amiloride inhibition of responses of rat single chorda tympani fibers to chemical and electrical tongue stimulations. *Brain Research, 451,* 319–325.

Norgren, R. (1974). Gustatory afferents to ventral forebrain. *Brain Research, 81,* 285–295.

Norgren, R. (1976). Taste pathways to hypothalamus and amygdala. *Journal of Comparative Neurology, 166,* 17–30.

Norgren, R. (1985). Taste and the autonomic nervous system. *Chemical Senses, 10,* 143–161.

Norgren, R., & Leonard, C. M. (1973). Ascending central gustatory pathways. *Journal of Comparative Neurology, 150,* 217–238.

Norgren, R., & Pfaffmann, C. (1975). The pontine taste area in the rat. *Brain Research, 91,* 99–117.

Norgren, R., & Wolf, G. (1975). Projections of thalamic gustatory and lingual areas in the rat. *Brain Research, 92,* 123–129.

Nowlis, G. H. (1977). From reflex to representation: Taste-elicited tongue movements in the human newborn. In J. M. Weiffenbach (Ed.), *Taste and development: The genesis of sweet preference* (pp. 190–203). Bethesda, MD: DHEW Publication No. (NIH) 77-1068.

Nowlis, G. H., & Frank, M. E. (1981). Quality coding in gustatory systems of rats and hamsters. In D. M. Norris (Ed.), *Perception of behavioral chemicals* (pp. 59–80). Amsterdam: Elsevier/North-Holland.

Nowlis, G. H., Frank, M. E., & Pfaffmann, C. (1980). Specificity of acquired aversions to taste qualities in hamsters and rats. *Journal of Comparative and Physiological Psychology, 94,* 932–942.

Oakley, B. (1967). Altered taste responses from cross-regenerated taste nerves in the rat. In T. Hayashi (Ed.), *Olfaction and taste II* (pp. 535–547). London: Pergamon.

Oakley, B. (1970). Reformation of taste buds by crossed sensory nerves in the rat's tongue. *Acta Physiologica Scandinavia, 79,* 88–94.

Oakley, B. (1988). Taste bud development in rat vallate and foliate papillae. In P. Hnik, T. Soukup, R. Vejsada, & J. Zelina (Eds.), *Mechanoreceptors* (pp. 17–22). New York: Plenum.

Oakley, B., Jones, L. B., & Kaliszewski, J. M. (1979). Taste responses of the gerbil IXth nerve. *Chemical Senses and Flavor, 4,* 79–87.

Ogawa, H., Sato, M., & Yamashita, S. (1968). Multiple sensitivity of chorda tympani fibres of the rat and hamster to gustatory and thermal stimuli. *Journal of Physiology (London), 199,* 223–240.

Ossebaard, C. A., & Smith, D. V. (1995). Effect of amiloride on the taste of NaCl, Na-gluconate, and KCl in humans: Implications for Na$^+$ receptor mechanisms. *Chemical Senses, 20,* 37–46.

Ossebaard, C. A., & Smith, D. V. (1996). Amiloride suppresses the sourness of NaCl and LiCl. *Physiology and Behavior, 60,* 1317–1322.

Parker, L. A., Maier, S., Rennie, M., & Crebolder, J. (1992). Morphine- and naltrexone-induced modification of palatability: Analysis by the taste reactivity test. *Behavioral Neuroscience, 106,* 999–1010.

Pfaffmann, C. (1941). Gustatory afferent impulses. *Journal of Cellular and Comparative Physiology, 17,* 243–258.

Pfaffmann, C. (1955). Gustatory nerve impulses in rat, cat and rabbit. *Journal of Neurophysiology, 18,* 429–440.

Pfaffmann, C. (1959). The afferent code for sensory quality. *American Psychologist, 14,* 226–232.

Pfaffmann, C. (1964). Taste, its sensory and motivating properties. *American Scientist, 52,* 187–206.

Pfaffmann, C. (1974). Specificity of the sweet receptors of the squirrel monkey. *Chemical Senses and Flavor, 1,* 61–67.

Pfaffmann, C., Erickson, R. P., Frommer, G. P., & Halpern, B. P. (1961). Gustatory discharges in the rat medulla and thalamus. In W. A. Rosenblith (Ed.), *Sensory communication* (pp. 455–473). New York: Wiley.

Pfaffmann, C., Frank, M., Bartoshuk, L. M., & Snell, T. C. (1976). Coding gustatory information in the squirrel monkey chorda tympani. In J. M. Sprague & A. N. Epstein (Eds.), *Progress in psychobiology and physiological psychology: Vol. 6* (pp. 1–27). New York: Academic Press.

Pritchard, T. (1991). The primate gustatory system. In T. V. Getchell, R. L. Doty, L. M. Bartoshuk, & J. B. Snow, Jr. (Eds.), *Smell and taste in health and disease* (pp. 109–125). New York: Raven.

Ramirez, I. (1991). Influence of experience on response to bitter taste. *Physiology & Behavior, 49,* 387–391.

Richter, C. P. (1956). Salt appetite in mammals: Its dependence on instinct and metabolism. In M. Autuori (Ed.), *L'Instinct dans de comportement des animaux et de l'homme* (pp. 171–176). Paris: Masson et Cie.

Richter, C. P., & Campbell, K. (1940). Taste thresholds and taste preferences of rats for five common sugars. *Journal of Nutrition, 20,* 31–46.

Rodieck, R. W., & Brenning, R. K. (1983). Retinal ganglion cells: Properties, types, genera, pathways and trans-species comparisons. *Brain, Behavior and Evolution, 23,* 121–164.

Rowe, M. H., & Stone, J. (1977). Naming of neurones: Classification and naming of cat retinal ganglion cells. *Brain Behavior and Evolution, 14,* 185–216.

Rozin, P., & Schiller, D. (1980). The nature and acquisition of a preference for chili pepper in humans. *Motivation and Emotion, 4,* 77–101.

Saper, C. B. (1982). Reciprocal parabrachial-cortical connections in the rat. *Brain Research, 242,* 33–40.

Saper, C. B., Swanson, L. W., & Cowan, W. M. (1979). An autoradiographic study of the efferent connections of the lateral hypothalamic area in the rat. *Journal of Comparative Neurology, 183,* 689–706.

Schiffman, S. S., & Erickson, R. (1980). The issue of primary tastes versus a taste continuum. *Neuroscience and Biobehavioral Reviews, 4,* 109–117.

Schwartz, G. J., & Grill, H. J. (1984). Relationships between taste reactivity and intake in the neurologically intact rat. *Chemical Senses, 9,* 249–272.

Scott, T. R., & Giza, B. K. (1990). Coding channels in the taste system of the rat. *Science, 249,* 1585–1587.

Scott, T. R., & Perrotto, R. S. (1980). Intensity coding in pontine taste area: Gustatory information is processed similarly throughout rat's brainstem. *Journal of Neurophysiology, 44,* 739–750.

Scott, T. R., & Plata-Salaman, R. (1991). Coding of taste quality. In T. V. Getchell, R. L. Doty, L. M. Bartoshuk, & J. B. Snow, Jr. (Eds.), *Smell and taste in health and disease* (pp. 345–368). New York: Raven.

Shingai, T. (1977). Ionic mechanisms of water receptors in the laryngeal mucosa of the rabbit. *Japanese Journal of Physiology, 27,* 27–42.

Shingai, T. (1980). Water fibers in the superior laryngeal nerve of the rat. *Japanese Journal of Physiology, 30,* 305–307.

Shingai, T., & Beidler, L. M. (1985). Response characteristics of three taste nerves in mice. *Brain Research, 335,* 245–249.

Shingai, T., Miyaoka, Y., & Shimada, K. (1988). Diuresis mediated by the superior laryngeal nerve in rats. *Physiology & Behavior, 44,* 431–433.

Shingai, T., & Shimada, K. (1976). Reflex swallowing by water and chemical substances applied in the oral cavity, pharynx, and larynx of the rabbit. *Japanese Journal of Physiology, 26,* 455–469.

Shipley, M. T. (1982). Insular cortex projections to the nucleus of the solitary tract and brainstem visceromotor nuclei in the mouse. *Brain Research Bulletin, 8,* 139–148.

Shipley, M. T., & Sanders, M. S. (1982). Special senses are really special: Evidence for a reciprocal, bilateral pathway between insular cortex and nucleus parabrachialis. *Brain Research Bulletin, 8,* 493–501.

Simon, S. A., Robb, R., & Schiffman, S. S. (1988). Transport pathways in rat lingual epithelium. *Pharmacology, Biochemistry and Behavior, 29,* 257–267.

Smith, D. V. (1985). The neural representation of gustatory quality. In M. J. Correia & A. Perrachio (Eds.), *Progress in clinical and biological research, contemporary sensory neurobiology, Vol. 176* (pp. 75–97). New York: Liss.

Smith, D. V., & Frank, M. E. (1993). Sensory coding by peripheral taste fibers. In S. A. Simon & S. D. Roper (Eds.), *Mechanisms of taste transduction* (pp. 295–338). Boca Raton, FL: CRC Press.

Smith, D. V., & Hanamori, T. (1991). Organization of gustatory sensitivities in hamster superior laryngeal nerve fibers. *Journal of Neurophysiology, 65,* 1098–1114.

Smith, D. V., Liu, H., & Vogt, M. B. (1994). Neural coding of aversive and appetitive gustatory stimuli: Interactions in the hamster brainstem. *Physiology & Behavior, 56,* 1189–1196.

Smith, D. V., Liu, H., & Vogt, M. B. (1996). Responses of gustatory cells in the nucleus of the solitary tract of the hamster after NaCl or amiloride adaptation. *Journal of Neurophysiology, 76,* 47–58.

Smith, D. V., & Theodore, R. M. (1984). Conditioned taste aversions: Generalizations to taste mixtures. *Physiology & Behavior, 32,* 983–989.

Smith, D. V., & Travers, J. B. (1979). A metric for the breadth of tuning of gustatory neurons. *Chemical Senses and Flavor, 4,* 215–229.

Smith, D. V., Travers, J. B., & Van Buskirk, R. L. (1979). Brainstem correlates of gustatory similarity in the hamster. *Brain Research Bulletin, 4,* 359–372.

Smith, D. V., Van Buskirk, R. L., Travers, J. B., & Bieber, S. L. (1983a). Gustatory neuron types in hamster brain stem. *Journal of Neurophysiology, 50,* 522–540.

Smith, D. V., Van Buskirk, R. L., Travers, J. B., & Bieber, S. L. (1983b). Coding of taste stimuli by hamster brain stem neurons. *Journal of Neurophysiology, 50,* 541–558.

Spector, A. C., & Grill, H. J. (1988). Differences in the taste quality of maltose and sucrose in rats: Issues involving the generalizations of conditioned taste aversions. *Chemical Senses, 13,* 95–113.

Spector, A. C., & Grill, H. J. (1992). Salt taste discrimination after bilateral section of the chorda tympani or glossopharyngeal nerves. *American Journal of Physiology, 263,* R169–R176.

Spector, A. C., Schwartz, G., & Grill, H. J. (1990). Chemospecific deficits in taste detection following selective gustatory deafferentation in rats. *American Journal of Physiology, 258,* R820–R826.

Stedman, H., Bradley, R., Mistretta, C., & Bradley, B. (1980). Chemosensitive responses from the cat epiglottis. *Chemical Senses, 5,* 233–245.

Steiner, J. E. (1977). Facial expressions of the neonate infant indicating the hedonics of food-related chemical stimuli. In J. M. Weiffenbach (Ed.), *Taste and development: The genesis of sweet preference* (pp. 173–189). Bethesda, MD: DHEW Publication No. (NIH) 77-1068.

Stellar, E. (1977). Sweet preference and hedonic experience. In J. M. Weiffenbach (Ed.), *Taste and development: The genesis of sweet preference* (pp. 363–373). Bethesda, MD: DHEW Publication No. (NIH) 77-1068.

Storey, A. T. (1968). A functional analysis of sensory units innervating epiglottis and larynx. *Experimental Neurology, 20,* 366–383.

Storey, A. T., & Johnson, P. (1975). Laryngeal water receptors initiating apnea in the lamb. *Experimental Neurology, 47,* 42–55.

Sweazey, R. D., & Bradley, R. M. (1986). Central connections of the lingual-tonsilar branch of the glossopharyngeal nerve and the superior laryngeal nerve in lamb. *Journal of Comparative Neurology, 245,* 471–482.

Sweazey, R. D., & Bradley, R. M. (1988). Responses of lamb nucleus of the solitary tract neurons to chemical stimulation of the epiglottis. *Brain Research, 439,* 195–210.

Sweazey, R. D., & Smith, D. V. (1987). Convergence onto hamster medullary taste neurons. *Brain Research, 408,* 173–184.

Travers, J. B. (1985). Organization and projections of the orofacial motor nuclei. In G. Paxinos (Ed.), *The rat nervous system: Vol. 2. Hindbrain and spinal cord* (pp. 111–128). Sydney: Academic Press.

Travers, J. B. Grill, H. J., & Norgren, R. (1987). The effects of glossopharyngeal and chorda tympani nerve cuts on the ingestion and rejection of sapid stimuli: An electromyographic analysis in the rat. *Behavioral Brain Research, 25,* 233–246.

Travers, J. B., & Norgren, R. (1983). Afferent projections to the oral motor nuclei in the rat. *Journal of Comparative Neurology, 220,* 280–298.

Travers, J. B., & Smith, D. V. (1979). Gustatory sensitivities in neurons of the hamster nucleus tractus solitarius. *Sensory Processes, 3,* 1–26.

Travers, S. P. (1993). Orosensory processing in neural systems of the nucleus of the solitary tract. In: S. A. Simon and S. D. Roper (Eds.), *Mechanisms of Taste Transduction.* Boca Raton, FL: CRC Press, pp. 339–394.

Travers, S. P., & Nicklas, K. (1990). Taste bud distribution in the rat pharynx and larynx. *Anatomical Record, 227,* 373–379.

Travers, S. P., & Norgren, R. (1991). Coding the sweet taste in the nucleus of the solitary tract: Differential roles for anterior tongue and nasoincisor duct gustatory receptors in the rat. *Journal of Neurophysiology, 65,* 1372–1380.

Travers, S. P., Pfaffmann, C., & Norgren, R. (1986). Convergence of lingual and palatal gustatory neural activity in the nucleus of the solitary tract. *Brain Research, 365,* 305–320.

Travers, S. P., & Smith, D. V. (1984). Responsiveness of neurons in the hamster parabrachial nuclei to taste mixtures. *Journal of General Physiology, 84,* 221–250.

Tyner, C. F. (1975). The naming of neurons: Applications of taxonomic theory to the study of cellular populations. *Brain, Behavior and Evolution, 12,* 75–96.

Van Buskirk, R. L., & Smith, D. V. (1981). Taste sensitivity of hamster parabrachial pontine neurons. *Journal of Neurophysiology, 45,* 144–171.

Van der Kooy, D., Koda, L. Y., McGinty, J. F., Gerfen, C. R., & Bloom, F. E. (1984). The organization of projections from the cortex, amygdala, and hypothalamus to the nucleus of the solitary tract in rat. *Journal of Comparative Neurology, 224,* 1–24.

Vogt, M. B., & Smith, D. V. (1993a). Responses of single hamster parabrachial neurons to binary taste mixtures: Mutual suppression between sucrose and QHCl. *Journal of Neurophysiology, 69,* 658–668.

Vogt, M. B., & Smith, D. V. (1993b). Responses of single hamster parabrachial neurons to binary taste mixtures of citric acid with sucrose or NaCl. *Journal of Neurophysiology, 70,* 1350–1363.

Wang, L., & Bradley, R. M. (1993). Influence of GABA on neurons of the gustatory zone of the rat nucleus of the solitary tract. *Brain Research, 616,* 144–153.

Wang, Y. T., & Bieger, D. (1991). Role of solitarial GABAergic mechanisms in control of swallowing. *American Journal of Physiology, 261,* R639–R646.

Whiteside, B. (1927). Nerve overlap in the gustatory apparatus of the rat. *Journal of Comparative Neurology, 44,* 363–377.

Wiggens, L. L., Frank, R. A., & Smith, D. V. (1989). Generalization of learned taste aversions in rabbits: Similarities among gustatory stimuli. *Chemical Senses, 14,* 103–119.

Woolston, D. C., & Erickson, R. P. (1979). Concept of neuron types in gustation in the rat. *Journal of Neurophysiology, 42,* 1390–1409.

Yamada, K. (1966). Gustatory and thermal responses in the glossopharyngeal nerve of the rat. *Japanese Journal of Physiology, 16,* 599–611.

Yamada, K. (1967). Gustatory and thermal responses in the glossopharyngeal nerve of the rabbit and cat. *Japanese Journal of Physiology, 17,* 94–110.

Yamaguchi, S. (1979). The umami taste. In J. C. Boudreau (Ed.), *Food taste chemistry* (pp. 33–51). Washington, DC: American Chemical Society.

Yamamoto, T. (1984). Taste responses of cortical neurons. *Progress in Neurobiology, 23,* 273–315.

Yamamoto, T., Matsuo, R., Kiyomitsu, Y., & Kitamura, R. (1988). Taste effects of 'umami' substances in hamsters as studied by electrophysiological and conditioned taste aversion techniques. *Brain Research, 451,* 147–162.

Yamamoto, T., & Yuyama, N. (1987). On a neural mechanism for cortical processing of taste quality in the rat. *Brain Research, 400,* 312–320.

Yamamoto, T., Yuyama, N., Kato, T., & Kawamura, Y. (1985). Gustatory responses of cortical neurons in rats. II. Information processing of taste quality. *Journal of Neurophysiology, 53,* 1356–1369.

Ye, Q., Heck, G. L., & DeSimone, J. A. (1993). Voltage dependence of the rat chorda tympani response to Na+ salts: Implications for the functional organization of taste receptor cells. *Journal of Neurophysiology, 70,* 167–178.

Yoshii, K., Kobatake, Y., & Kurihara, K. (1981). Selective enhancement and suppression of frog gustatory responses to amino acids. *Journal of General Physiology, 77,* 373–385.

Yoshii, K., Yoshii, C., Kobatake, Y., & Kurihara, K. (1982). High sensitivity of *Xenopus* gustatory receptors to amino acids and bitter substances. *American Journal of Physiology, 243,* R42–R48.

Zalewski, A. A. (1969). Role of nerve and epithelium in the regulation of alkaline phosphatase activity in gustatory papillae. *Experimental Neurology, 23,* 18–28.

CHAPTER 3

Psychophysics of Taste

Bruce P. Halpern

I. PROLOGUE

A. The Nature of This Chapter

This chapter focuses on five aspects of human taste psychophysics that I believe to be not only quite fundamental but also in need of discussion. Each aspect is framed as a question. Some topics, such as the identification and specification of taste stimuli, or the appropriate terms to be used to describe perceived taste, have received what I consider to be excessively facile and circumscribed answers. Another aspect, the general theory underlying taste perception, is settled for many chemosensory workers, but quite prematurely in my judgment. A fourth topic, exploring taste perception as an ongoing process over time, presents a concept familiar to students of visual, auditory, and somesthetic perception, and to food scientists, but unknown by many other chemosensory investigators. Finally, the relevance of gustatory absolute thresholds to normal taste judgments is discussed. An attempt is made to relate each topic to taste-dependent decisions that occur under conditions of normal eating and drinking.

This prologue contains three brief sketches of human chemosensory-dependent behaviors that are found in relatively ordinary, normal situations. Each description contains, or is followed by, a series of questions related to the behaviors. The purpose of these sketches is to offer a small sample of the types of common yet important occurrences for which chemosensory perception is an indispensable factor.

Tasting and Smelling

77

Questions are offered to suggest some of the many points of discussion and inquiry that are raised by each example of chemosensory perception.

Although taste psychophysics would be involved in the study of every instance depicted, the questions, and the events from which they derive, are very much broader in scope than this chapter. Then why present them? My hope is that they will serve to suggest that chemosensory perception in general, and taste perception in particular, is sufficiently interesting, important, and complex to be worthy of serious study and thought by all students of perception and sensory systems. Psychophysical concepts, theory, and assumptions deeply influence the nature of, and outcomes from, any study of taste perception.

B. Three Very Short Stories

1. Taste as the Crucial Element in Ingestion Decisions

You are in a beautiful dining room. The lighting is perfect. Appropriate, pleasant music can be heard. Food and drink are served to you on a very attractive plate and in a handsome glass. It all looks wonderful. You place some food into your mouth and within a second or two say, or at least think, "This doesn't taste right." You take at most one more bite, and then choose to eat no more.

How did this happen? Everything was going so well until you put the food or beverage into your mouth. An experience that seemed very positive became unsatisfactory and disappointing.

What aspects of the food underlay your decision to reject it? Can these aspects be predictably altered by varying the chemical composition of the food or the temperature at which it is eaten?

Would all human adults respond in the same manner that you did? If not, would differences in age, health, genetic heritage, developmental conditions before or after birth, or individual life-history be important?

2. Taste Perception as a Spatial–Temporal Pattern

Your favorite beer is not available. You accept a suggested alternative. The beer is served in an ordinary glass. It looks rather like other beers. Because the beverage is unfamiliar, close attention is paid when it is drunk. You quickly notice that the temperature of the beer is "correct," and then, over several seconds, experience complex perceptual patterns, apparently originating in your mouth, that change several times both before and after you swallow. You think "that's what beer should be like," and say "That's good."

What accounts for the complex, changing, mouth-related patterns that you experienced? Is it possible to measure them? Can the patterns be usefully analyzed into separate components that are dependent on certain chemicals or groups of chemicals in the beverage? Does the temperature of the beverage affect the patterns that you perceived?

Are aspects of these patterns connected to particular parts of the mouth, or to specific sensory organs in and around the mouth? Do areas of the mouth, or oral sensory organs, differ in sensitivity or responsiveness to certain chemical, thermal, or mechanical events? Would abnormal conditions in one or more of these sensory organs change the patterns in predictable ways? What neural pathways connect the areas of the mouth, or the relevant sensory organs, to the human central nervous system? Would the same pathways and sensory organs be involved in all mammals?

Do known regions of the human nervous system relate to the complex patterns that you experienced? In your brain, might the activity of certain regions change in predictable ways when some chemicals or groups of chemicals are in or around the mouth? If so, would the same changes be expected in all humans, or perhaps in all mammals?

3. Perception of Mixtures

You are cooking a dish that you have never prepared before. Two friends arrive, and you ask each of them to taste the food at its present stage. Both do so. One remarks, "It needs some oregano," and the other says "Try adding a little lemon juice." You divide the food into four portions. Oregano is added to one portion, lemon juice to the second, and both the third portion. The fourth is kept unchanged. Now all three of you taste the four portions.

Will the portions to which oregano or lemon juice were added necessarily now taste as lemon juice or oregano would taste if alone? Should the portion to which both ingredients were added have a more intense taste than any of the other portions?

Can the taste of the four portions, as reported by each of you, be objectively and quantitatively measured? If so, is particular training or practice required? Are such measurements useful in understanding the perceptual, cognitive, or sensory nature of the experiences evoked by the food? If you place some of the four portions in fresh containers such that only you know which is which, would your friends be able to identify them correctly?

Another friend arrives and also samples everything, but without being told anything about the preparation conditions. For each sample, you ask "What's in it?" Would you expect correct identification of some or all components? Suppose your friend said the portion with lemon juice had a "citrus taste," or was "slightly sour," or was "tart," or was "bitter," or was "like humus"? Should any of those descriptions be accepted as correct for the lemon juice component? If instead you had first given your friend a taste of one of the components and then had asked your friend to tell you if any of the four portions contained that component, would they easily identify the corresponding portion of the food? Might foods made with many components, or those that are composed of a large amount of some ingredients and rather small quantities of others, differ from those with few components or similar quantities of each component? Should your expectations be different depending on whether the new arrival is a man or a woman?

The dish, which is a little more than moderately spiced, is being prepared for a dinner at which there will be young children in the 2- to 4-year-old range, teenagers, and adults ranging from age 22 to age 93. Is the food likely to taste the same to individuals throughout this age range? If not, are there known changes with development or age that suggest what the differences in experienced taste will be? If you had a way in which all ages at the dinner could both describe this food and indicate how much they liked or disliked it, would you expect that descriptions would be age independent, whereas preferences would not?

II. INTRODUCTION

A. Taste and Human Behavior

The three scenarios just given present situations that are quite common in human experience. With a little modification of the specifics, they could fit many events in a wide range of cultures. In each instance, taste-dependent decisions and actions occur. These behaviors are, I believe, both central parts of human life and matters of substantial interest to human societies. One reason is a set of cultural prescriptions and prohibitions that define what may or should be eaten or drunk, and what ought to not be ingested. Avoidance of specific food or beverage samples that are likely to produce illness is a second reason. The pleasure associated with both the preparation and the consumption of food or drink, even when one may have little or no metabolic need for eating or drinking at that time, is a third significant factor.

Each scenario was followed by several questions. Some questions addressed the significance of tasting in human behavior; the measurability of taste-dependent experiences; influences of general factors such as age, disease, heredity, and developmental conditions on tasting; and the role of individual experiences. Other questions asked whether particular chemicals or groups of chemicals in foods or beverages yielded predictable tastes, whether different tastes or kinds of tasting were related to certain regions of the mouth, and if descriptions of taste, and liking or disliking what one tastes, are separable. Potential problems were noted concerning which taste descriptors were "correct," and the dynamic, changing nature of perceived taste was emphasized. The possibility was raised that multicomponent foods might be perceived as being quite different from their components, and that sensory interactions might occur between effects of the thermal, mechanical, and chemical properties of foods and beverages. Finally, relationships between tasting and nervous system structure and function were raised. Underlying themes were the suggestions that human taste experiences can be considered patterns that occur over time, that the relevant temporal parameter is seconds or perhaps fractions of a second, and that measurements of chemosensory perception are likely to be possible but perhaps challenging.

A comprehensive work on taste would provide serious treatments of all these questions; many are approached in the chapters of this volume. This chapter on human taste psychophysics has a narrow purview: an inspection of some of the concepts and assumptions that underlie psychophysical measurement of taste-dependent judgments, with an examination of certain relevant data. The psychophysical measurements allow formal exploration of relationships between taste-dependent judgments and the chemical and physical events that elicited the judgments. Appropriate selection of psychophysical gustatory measurement procedures, and interpretation of results, require an understanding of what taste processes are actually being accessed, knowledge of why those measurements and not others are being made, and a critical appreciation of the conceptual framework within which one is operating. Identification, discussion, and analysis of a few of these concerns are goals of this chapter.

On the other hand, many important taste-related questions will not be addressed here. The significance of tasting in human behavior, although a necessary assumption to justify this chapter, must be treated separately (see Ch. 6), as must study of human food choice and eating behavior per se (e.g., Meiselman, 1988, 1992a, 1992b, 1993). In similar fashion, although the data produced by the psychophysical methods discussed in this chapter are essential for identifying and evaluating proposed physiological mechanisms of tasting (Halpern, 1986a, 1991), physiology is beyond the scope of this chapter. Taste physiology can be found in Ch. 1 and 2. Both gustatory physiology and anatomy are addressed in a number of other current publications (e.g., Finger & Silver, 1987; Getchell, Doty, Bartoshuk, & Snow, 1991; Kuznicki & Cardello, 1986; Simon & Roper, 1993).

Also largely absent from this chapter are practical instructions on the application of psychophysical procedures to the study of taste, guidance on the statistical or numerical analysis of gustatory psychophysical observations, and tabulations of gustatory thresholds or descriptions for various pure chemicals in aqueous solution. Many excellent sources for this information already exist. Examples include chapters both in general treatises (Bartoshuk, 1988; Bartoshuk & Marks, 1986; Bolanowski & Gescheider, 1991; McBurney, 1984) and in volumes devoted to chemosensory function (Getchell et al., 1991; Lawless & Klein, 1991; Meilgaard, Civille, & Carr, 1991; Meiselman & Rivlin, 1986; Moskowitz, 1984; O'Mahony, 1986; Piggott, 1988; Stone & Sidel, 1993). Earlier publications can also be useful (e.g., Amerine, Pangborn, & Roessler, 1965; Kling & Riggs, 1971; Pfaffmann, 1959).

B. Natural Complexity and Laboratory Reductions

Tensions often exist between the desire of a researcher to study events as they actually occur in the world, and the need to make repeated, systematic observations under known and controlled conditions. Quantitative, highly controlled experiments may require that only one variable be manipulated, and that this variable be one that represents a single, well-described physical continuum. The latter circumstance represents a common pattern for research done in physics laboratories. Physics is of course the namesake of, and frequently also the model for, psychophysics (Boring, 1950; Watson, 1973).

Research in psychophysics commonly proceeds by reducing the complexity of natural seeing, hearing, touching, eating, or drinking to a single variable that is studied under controlled conditions. This approach has yielded substantial amounts of valuable data. Sometimes, however, the desire to obtain general and elegant relationships between physical variables and behavioral responses, relationships that are comparable to the "laws" of physics, can result in a loss of relevance to natural, real-world circumstances. Limitations of this sort apply to the laws of physics as well as to those of psychophysics. For example, the generalization by Galileo that the velocity of falling bodies is independent of their mass, although a major creative insight, could result in serious problems if a manufacturer of parachutes concluded that a parachute that was adequate to slow the descent of a truckload of dry hay would therefore be suitable for the same truck filled with heavy equipment (Simon, 1993).

The general problem is that transitions from the richness of natural circumstances to the restrictions of laboratory investigations are often accompanied by seriously diminished pertinence to normal life (see Meiselman, 1992a, 1992b). Unfortunately, this untoward consequence is sometimes unavoidable. Maintaining awareness of the hazard may be at least a partial remedy. It can temper one's readiness to generalize from unique and narrow situations to broad conclusions.

Nonetheless, prevention of problems is better than remedial treatment. When possible, laboratory studies should be designed to include essential aspects of the normal circumstances. One approach is careful observation of the original events that one wishes to understand. The goal is to identify crucial boundary conditions that should be included in controlled investigations.

To accomplish this aim when studying the psychophysics of human taste, utilization of chemical compounds that occur in foods as stimuli is beneficial. However, in isolation it is not sufficient, even if the food-derived chemicals are presented in the groupings and ratios that occur in foods. Our taste perception depends on many events in the mouth. Intraoral thermal, mechanical, chemical, or fluid dynamical conditions, and the time course of these events, could be important (Gibson, 1966). Some of the fundamental parameters may be the rates at which humans transfer food or drink into and out of the mouth, the volumes ingested, and the texture, temperature, chemical composition, and intraoral solubility of foods and beverages, all as they occur when the foods or beverages are eaten or drunk.

Taking account of many different factors and conditions, with small differences in the combinations or patterns expected to have large effects, is not very unusual in biology. With reference to physical sciences, it is perhaps more akin to chemistry than to physics (Hoffmann & Torrence, 1993). To this extent at least, description of taste as a chemical sense may be quite appropriate. Human taste stimulation should be understood to include all the relevant intraoral circumstances in which the chemicals occur, rather than just the chemicals themselves.

III. SOME FUNDAMENTAL QUESTIONS OF TASTE PSYCHOPHYSICS

A. What Are Taste Stimuli?

1. A Preliminary Example: Visual Stimuli

With some sensory systems, straightforward descriptions of the typical environmental energies that activate the system have long been available. For example, we say that light is the stimulus for our visual system. This is an abbreviated way of stating that the typical environmental energy for vision is "Electromagnetic radiation that has a wavelength in the range of about 3,900 to about 7,700 angstroms and that may be perceived by the unaided, normal human eye" (American Heritage Electronic Dictionary, 1992). An important part of this definition is the assertion that vision is dependent on a small and fully specifiable portion of a single physical continuum.

I qualified the definition for light in advance by using "typical" in the sentences preceding it. This was done because for most sensory systems many different types of environmental events can active the system, but only a limited subset is plausible for natural or generally encountered conditions. Thus, we know that cosmic rays, electric currents, and mechanical events can also stimulate the visual system (see Plattig, 1991), but are not believed commonly to be effective in the usual terrestrial environments. I am in effect resurrecting the ancient discussions that led in the late 19th century to the concepts of inadequate and adequate stimuli (Boring, 1942; Woodworth & Schlosberg, 1954), with the latter related to my use of "typical." This is not an attempt to produce highly selective, narrowly tuned perceptual systems by asserting that some of the stimulus events of normal living are irrelevant. Rather, I am acknowledging that because all receptor and perceptual selectivity is relative rather than absolute, any system will occasionally be activated by intense natural stimuli that are not very meaningful for that system—for example, visual experience induced by a blow to the eye. In addition, events created in the laboratory, such as electric currents, can stimulate vision and most other system systems. This can be useful for biophysical research or clinical purposes (see Getchell et al., 1991; Meiselman & Rivlin, 1986), but is usually not germane to an understanding of perception.

Relatively simple operations underlie the preceding definition of typical stimuli for vision. The stated electromagnetic radiation is both necessary and sufficient for vision. Other portions of the electromagnetic spectrum are largely or totally ineffective; patterned vision does not result from mechanical events or electrical currents grossly applied to the eye.

2. Taste Stimuli

a. Chemical Stimuli

When tests of necessity and sufficiency comparable to those used for visual stimuli are applied to taste, no single physical or chemical dimension is revealed (Lawless, 1987; Scott & Plata-Salaman, 1991). With regard to chemicals per se, taste–active substances in foods and beverages range from hydrogen, with an atomic mass of 1, through metals such as sodium, with atomic mass in the twenties (actually hydrogen or sodium ions), to proteins with molecular weights between 10 and 30,000 kDA [e.g., curculin, mabinlins, monellin, thaumatin (Cagan, Brand, Morris, & Morris, 1978; Inglett, 1978; Kurihara, 1992, 1994; Van der Wel, 1972, 1993; Yamashita et al., 1990)]. The taste–active compounds in foods include alcohols, alkaloids such as the quinoline alkaloids, amino acids, carbohydrates, flavonoids, metals, peptides, proteins, triterpenes, a wide range of other organic compounds, and inorganic and organic salts and acids (Boudreau, 1979; Fennema, 1985; Fuke & Konosu, 1991; Kato, Rhue, & Nishimura, 1989; Robins, Rhodes, Parr, & Walton, 1990; Rouseff, 1990; Schiffman & Erickson, 1980; Schiffman, Hopfinger, & Mazur, 1986).

The preceding extensive list of chemicals that serve as taste stimuli may be surprising if one is accustomed to statements that there are only four types of taste stimuli. Four chemicals, often the metallic chloride salt NaCl, the carbohydrate saccharide sucrose, an alkaloid derivative such as a quinine salt, and an acid, customarily the inorganic acid HCl, are frequently specified as the basic, primary (Kurihara, 1987; Zapsalis & Beck, 1985), fundamental (Rouseff, 1990), or perhaps less expansively, prototypical taste stimuli. This list of four putative primary taste stimuli actually derives from a theoretical approach rather than from empirical data (see O'Mahony & Ishii, 1987). After the present analysis of taste stimuli, I will discuss attempts to provide a theory for taste perception.

By elaborating the discrete chemical and other events (see later) that can be identified as effective stimuli during our contact with foodstuffs, I do not mean to imply that a single specified chemical, perhaps at a particular temperature, represents a natural or ecologically valid taste stimulus. Our food and beverage selections generally are not composed of one or two items from the list of taste stimulus chemicals given above. Instead, we eat and drink complex mixtures made up of many chemicals. Often, these mixtures are present in an intricate physical array. It is such food and beverage systems that are the real-world taste stimuli (e.g., Gillette, 1985; Redlinger & Setser, 1987).

Preference measures are frequently applied to these normally consumed food and drink items (see Ch. 6), but nonhedonic psychophysical analyses of food and beverage systems that yield information on taste descriptions, taste intensity, or thresholds are uncommon outside the food science tradition (O'Mahony, 1990; Pangborn, 1981; Trant & Pangborn, 1983), although not unknown. One reason may be that predetermined, highly restrictive, and relatively arbitrary taste descriptors or intensity categories can be ineffective when studying beverages or foods that humans actually drink or eat (e.g., Schiffman & Sattely-Miller, 1994).

Difficulties in applying taste psychophysics to foods and beverages can be reduced by using food components, for example, soup bases, to align or operationally define certain taste concepts (O'Mahony, 1991). Basically, this approach presents an exemplar for a taste concept, and hopes that the resultant chemosensory concept will not be excessively fuzzy. If delimitation is required, negative exemplars are added (Ishii and O'Mahony, 1987, 1991). Within food science, this methodology is often used to enable several judges to apply descriptors in a similar manner (e.g., Powers, 1988). However, it can also be used to allow gustatory decisions to be made in the total absence of linguistic labels. Under these conditions, judges show rapid and consistent taste-dependent acceptance and rejection (Delconte, Kelling, & Halpern, 1992).

b. Nonchemical Stimuli for Taste

Environmental events other than chemicals are also effective taste stimuli. Neurophysiological studies in nonhuman vertebrates demonstrate that moderate temperature changes and nontraumatic mechanical deformation within the mouth or on the tongue can change the neural activity of individual neurons that innervate taste receptor cells (see Andersen, 1970; Faurion, 1987b; Smith & Frank, 1993). There is no reason to believe that this is not also true for humans. Consequently, definition of the taste perceptual system as a chemical sense is too narrow. The thermal, mechanical, and chemical events associated with eating and drinking, and with other uses of the tongue and mouth, can all modulate the activity of our taste receptor cells and the neurons that connect these cells to the brain. In effect, this means that the temperature changes, mechanical deformations, and chemicals reaching our taste buds during the ingestion, chewing, and swallowing of food and drink may all increase, decrease, or change the temporal distribution of the action potential traffic in the neurons that innervate taste buds. Put more simply, they can all function as taste stimuli (Gibson, 1966).

Inclusion of many chemical, thermal, and mechanical events as potential taste stimuli does not deny that some of these taste stimuli may also provide somatosensory input (Breslin, Gilmore, Beauchamp, & Green, 1993; Green & Lawless, 1991; Hettinger & Frank, 1992; Prescott, Allen, & Stephens, 1993). In this case, the stimuli might affect both the trigeminal system, which is another sensory system innervating the tongue and other structures of the oral cavity as well as separate regions of the head (Silver & Finger, 1991; Simon & Wang, 1993), and other neural systems found in the mouth. Under normal circumstances, trigeminal and taste stimulation will both take place during eating and drinking, contributing to the overall experience (Gibson, 1966). Selective trigeminal responses to chemical, thermal, and mechanical stimulation occur to stimulation of the tongue, although the required stimulus intensities are typically substantially greater than those needed for "taste" responses (Wang, Erickson, & Simon, 1993). Additional distinctions can be made under controlled conditions. The time intervals between stimulus onset and trigeminal responses are relatively long, whereas the effectiveness of various chemicals as trigeminal stimuli is proportional to the ease with which they dissolve in lipids (Simon & Wang, 1993). The longer reaction times, more gradual perceived onsets, and correlation with lipid solubility may all be related to the location of trigeminal receptor structures further below the surface of the skin than are taste buds, and the lack of accessibility through openings (pores) that do permit relatively direct contact between intraoral fluids and extensions of the receptors cells of taste buds (Green & Lawless, 1991; Silver and Finger, 1991). In addition, repeated or sustained exposure to chemicals that are effective trigeminal stimuli has cumulative and sustained consequences. The term *oral chemesthesis* has recently been applied to these gradual and relatively long-term alterations in perceived intensity of certain chemical stimuli.

3. Taste, Flavor, Touch, and Olfaction

Most students of sensory function limit human taste perception to events initiated by stimulation of certain structures of the oral cavity. Although I do not quarrel with this limitation, I do include within the set of taste stimuli at least some of the thermal and mechanical concomitants of eating, drinking, licking, swallowing, etc., as suggested above. Even with this inclusion, many of the physical and perhaps some of the chemical changes that occur during our biting, licking, chewing, and suction drinking of foods and other items remain to be considered (see Lawless, 1987). Some of these stimulus events find a place within the rather broad perceptual concept of *flavor*. "Flavor is the complex of sensations through which the presence and identity of foods and beverages in the mouth are identified" (Crocker, 1945). For current food chemistry, food science (Powers, 1988; Zapsalis & Beck, 1985), and perceptual analyses, flavor is stipulated to include the taste, odor, mouth feel, visual, and auditory aspects of foods: "Generally, the term 'flavor' has evolved to a usage that implies an overall integrated perception of all the contributing senses (smell, taste, sight, feeling, and sound) at the time of food consumption" (Lindsay, 1985), and "Taste in the everyday, as well as the Gibsonian, sense is an active perceptual system that makes use of many separate senses, including taste, smell, the common chemical sense, temperature, touch, vision, and hearing ('They taste as good as they crunch'). All of these contribute to *flavor*" (McBurney, 1986). Flavor thus subsumes taste, oral chemesthesis including nontaste input via the trigeminal system, stimulation through the olfactory system, and, if it is functional in humans, perhaps activation of the vomeronasal system (Monti-Bloch & Grosser, 1991). When flavor is taken as a full compendium of perceptual aspects of eating and drinking (Gibson, 1966), it also includes the acoustic aspects of foods, input from the teeth and jaws during biting and chewing (Dubner, Sessle, & Storey, 1978), and visual impressions. Fortunately, all but taste are excluded from consideration in this chapter. However, it should be recognized that a realistic discussion of perception or cognition during eating and drinking requires consideration of all the relevant sensory and perceptual systems or inputs just listed, and their function as a cognitive entity (Lindsay, 1985; McBurney, 1984, 1986).

B. Theory: Is Taste Perception Four Discrete, Independent Processes or a Pattern of Overlapping Events?

1. Underlying Models

The richness of the stimuli that evoke taste perceptions and the complexity of this perceptual system during normal taste-dependent decisions have been a challenge to theorists (Andersen, 1970). Other, more deeply analyzed perceptual systems have been drawn on to provide templates for functional explanations. Two general approaches have been used to describe the nature of the process or processes involved in taste perception (see DeSimone, 1991; Smith & Frank, 1993). One of these approaches, the basic tastes/taste primaries theory, regularly uses the auditory system as a conceptual model (Kroeze, 1990; McBurney, 1986; Scott and Plata-Salaman, 1991). Two aspects of vertebrate auditory systems are emphasized: first, the physical separation of sound frequencies in peripheral auditory organs, that is, "the cochlea acts as if it contained a series of bandpass filters with continuously overlapping center frequencies," and second, the perceptual ability of some human observers to identify the constituent sound frequencies of a complex sound if the component frequencies are sufficiently separated (Backus, 1977; Gulick, Gescheider, & Frisina, 1989).

   From anatomical and physiological perspectives, the approximately 3000 spatially ordered inner hair cells of the human cochlea, which are the mechanoreceptor cells that drive the sensory nerve connecting the cochlea to the central nervous system, represent the elements that translate electromechanical separation of sound frequencies into afferent bioelectric activity (Pickles, 1988; Gulick et al., 1989). Given this large number of sound transducers, one might expect that a mammalian auditory system-based model, such as the basic tastes theory is considered to be, would predict hundreds if not thousands of different taste receptor types, and a great many different perceptual taste categories. Furthermore, psychophysical data indicate that in the frequency range from 100 to about 10,000 Hz, many humans can discriminate two tones if the frequencies differ by at least 5 Hz (Backus, 1977), thus suggesting the possibility of about 2000 pairwise discriminations. Again, one might expect the basic tastes theory, which uses as an analogy auditory data, to propose very many discriminable tastes.

   Our expectations would be incorrect. The basic or primary tastes theory proposes that very few distinguishable tastes exist, and that each is quite separate from the others. The number is most commonly taken to be four, corresponding to four distinct and rather independent receptor types (Bartoshuk, 1988; O'Mahony & Ishii, 1987).

How can a theory that conceptually derives from mammalian audition produce such an outcome? Not easily! Three rather large steps are required. First, from the complexity of the mammalian auditory system's physiology and anatomy a single idea is extracted—the separation of environmental stimuli as a basic property of a receptor organ. Second, because the thousands of hair cells and primary afferent neurons of mammalian auditory systems show increasingly broad tuning and considerable overlap of responsiveness as sound pressure level is increased (Pickles, 1988; Gulick et al., 1989), a very much simpler auditory system, such as that found in some moths and butterflies (see Scoble, 1992), is implicitly envisioned. Finally, instead of human auditory perception in which musical tones can routinely combine to yield new percepts, which in turn can sometimes be analyzed into the components by a trained expert, we are given a taste perceptual system with four readily distinguishable percepts for which synthesis does not occur. I shall discuss the basic tastes theory in more detail after first introducing its well-known but usually rejected competitor, the across-fiber or pattern theory of taste perception.

The other major theoretical direction for taste theory takes vertebrate color vision as a model system for taste perception (Erickson, 1963, 1977; Erickson & Covey, 1980; Faurion, 1987a; Schiffman & Erickson, 1980; Scott & Plata-Salaman, 1991). In common with the basic tastes theory, two aspects of a reference sensory system are emphasized for purposes of analogy by the pattern model of taste: First, there is the broad responsiveness of the receptor elements, now cone photoreceptor cells of the retina, each of which can be activated by large (in mammals about 50 to 70% of the visual spectrum), sequential (along the 390- to 770-nm region of the electromagnetic continuum), overlapping sets of wavelengths, as stimulus intensity is increased above threshold levels (Dowling, 1987). Second, there is the necessity for the visual system to utilize relative response across the different types of cones and the retinal neurons whose activity the cones modulate. This is the case because any one type of cone is unable to have its output *uniquely* controlled by a particular wavelength of light at any intensity above absolute threshold (Gouras & Zrenner, 1981). Thus, receptor elements and neural and perceptual processes with broad and overlapping sensitivities are emphasized. Relative activity of the constituent processes, and the overall patterns produced by such activity, are considered crucial for the pattern/across-fiber/population model of taste (Faurion, 1987b; Oakley, 1986; McBurney, 1986; Schiffman & Erickson, 1993; Smith, 1985; Smith & Frank, 1993).

Some other aspects of vertebrate color vision are generally ignored by the pattern theory of taste. An important one is that very few types of cones exist. The number rarely exceeds five in vertebrates, and is limited to three types in mammals (Mollon, 1982). This could suggest that only three or four types of taste receptors should be found, but a larger, although unspecified, set is implied by advocates of the population/pattern theory of taste (Erickson & Covey, 1980; Erickson, Priolo, Warwick, & Schiffman, 1990; Schiffman & Erickson, 1980, 1993). Another problem area for this taste theory in reference to its color vision analogy is the standard observation of visual psychophysics that appropriate discrete wavelengths of visible light (usually a short, middle, and a long wavelength), if overlapped in the visual field, will produce a percept for which the component wavelengths cannot be veridically identified, and which itself cannot be discriminated from another visual mixture produced by three different wavelengths selected from the same general regions of the visual spectrum (Mollon, 1982). If taste perception held to this color vision principle of total synthesis, components of gustatory mixtures could never be veridically identified (O'Mahony, Rothman, Ellison, Shaw, & Buteau, 1990), and complete matching of a taste percept produced by one mixture could be routinely done using a different mixture.

The literature on perception of taste mixtures is complex and confusing. However, neither a total inability to identify components of a mixture nor uniform and consistent success in specifying the ingredients has been found; a ready ability to match a mixture with another mixture or a solution of a single chemical has been rarely reported (see Bartoshuk & Gent, 1985; Burgard & Kuznicki, 1990; Erickson & Covey, 1980; Erickson et al., 1990; O'Mahony & Buteau, 1982; O'Mahony, Atassi-Sheldon, Rothman, & Murphy-Ellison, 1983; Schiffman & Erickson, 1980, 1993); see Section III,B,3,b.

Perhaps two disclaimers should be added. First, for both visual and gustatory mixtures, one could learn to report more or less correct components from the perception of the mixture and a previously acquired knowledge of likely constituents (e.g., a blue, a green, and a red light for a visual perception of "white"; lemon juice, egg yolk, and oil for a gustatory perception of "mayonnaise"). This indicates that more or less veridical reporting of components is not sufficient to demonstrate an ability to analyze or decompose a mixture unless there is independent evidence that the judge has not been, in effect, previously tutored as to the ingredients. Second, metameric color matching, in which three visual stimuli widely separated in wavelength are used to match a fourth light by varying the intensity and position of three of them, is not a straightforward process (Levine & Shefner, 1991). Specifically, varying position means that one or more of the set of three may have to be added to the fourth light for which matching is being done, in order to make the visual match: "negative scalars are usually required for one or more of the components, meaning that those components must be admixed to the side of the field containing D" (Pugh, 1988), where D is the light to be matched.

Both the basic tastes and the pattern theories of taste perception have roots that can be usefully followed into the past. Historical analyses of these theories are available in several sources (e.g., Bartoshuk & Gent, 1985; McBurney, 1986; Schiffman & Erickson, 1980, 1993).

2. Analysis, Synthesis, or Fusion?

Connections are often made between the taste primaries/basic tastes model of taste perception and an analytic view of sensory and perceptual systems (Bartoshuk, 1988; Bartoshuk & Gent, 1985; Burgard & Kuznicki, 1990; Kroeze, 1990), whereas

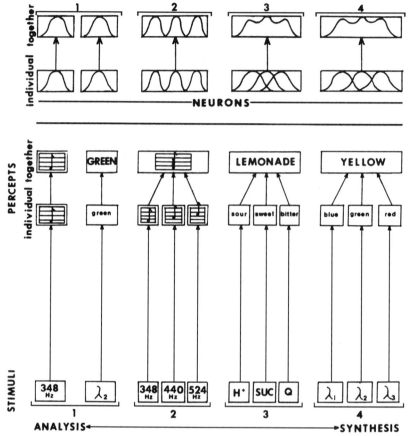

FIGURE 1   Pictorial definitions of analysis and synthesis at several different levels of investigation. Distinguishable levels are separated vertically. The lowest level of the figure illustrates several different types of physical stimuli: pure tones, designated by integers indicating sound frequency in hertz; wavelengths of light (λ); and aqueous solutions of single molecular species (H^+, SUC, Q). For the wavelengths, subscripts 1, 2, and 3 denote light perceived as blue, green, and red, respectively. For the chemicals, H^+ indicates a solution with a hydrogen ion concentration such that it is perceived as primarily sour; SUC is a sucrose solution that is perceived as primarily sweet; Q is a solution of a quinine salt that is perceived as primarily bitter. The middle levels of the figure, located below the horizontal solid line and connected to the stimulus level with vertical lines, represent the percepts. They represent perception of the stimuli by normal adult humans, either when one stimulus is presented by itself ("individual"), or when two or three stimuli are presented as a mixture ("together"). The upper levels of the figure depict central nervous system neural responses to individual or mixed stimuli. Analysis, fusion, and synthesis are presented from left to right. On the far left (panel 1) a completely analytical situation is shown. Here, a 348-Hz tone and a visual stimulus that is perceived as green are presented either separately or mixed. When presented together, the two stimuli are nonetheless perceived as fully separate, with no novel percept appearing, and produce quite separate neural responses. The relevant neurons are responsive either to the sound or to the light, with no overlapping sensitivity. On the far right (panels 3 and 4) fully synthetic processes are illustrated. For panel 3, solutions that are perceived as primarily sour, sweet, or bitter when presented separately will yield a new and different percept, lemonade, when presented as a

the population/pattern model of taste is frequently considered to require a synthetic approach to sensory and perceptual events (Erickson, 1977; Erickson & Covey, 1980). The concepts of analysis and synthesis, although sometimes presented as deductions based on the characteristics of particular receptor organ structures or neural interconnections, actually refer to the apparent nature of mental processes, as revealed by introspection (Sekuler & Blake, 1985). "In an analytic sense, two stimuli that are presented in a mixture maintain their individual qualities of sensation without loss or the introduction of a new quality. In a synthetic sense, on the other hand, two stimuli that are mixed lose their individual qualities and a new quality arises" (McBurney, 1986).

As just suggested, it is widely believed that taste must be understood as either a synthetic or an analytic system (Figure 1). However, these conflicting strong claims, instantiated as the assertion that the perception of taste mixtures must totally conform to outcomes now associated with the pattern or the basic tastes theories, do not appear to be logically required. This can be illustrated with an example from auditory perception. Sound perception is often used to illustrate a system in which individual stimulus elements, specifically sound frequencies or the sounds produced by individual musical instruments, can sometimes be recognized when a "mixture" of frequencies, musical instruments, or notes is heard (Moore, 1982). This can be taken as an example of analysis. In addition, however, the tonal pattern produced either at one moment or over time by the acoustic mixture may be a percept clearly different from the individual elements. We now recognize synthesis. Indeed, with both music and speech, the essential nature of the acoustic mixture as a separate pattern or percept can be preserved even when all the constituent sounds are changed. For example, all speech frequencies can be made higher or lower, or music normally performed on a piano or a clarinet can be played with a drum or a harmonica, but language communication, or the tune, is preserved. If the auditory examples just given are valid for taste perception, one might expect that taste mix-

---

mixture. The individual components cannot be identified in either the "together" percept or the corresponding neural response. In panel 4, a comparable situation is seen for three lights. Alone, one light is perceived as blue, another as green, the third as red. When presented as an overlapping mixture, a new and different percept, yellow, occurs, with no perceptual indication of the constituent wavelengths. For both synthetic processes, the relevant taste or visual neurons have relatively broad, overlapping sensitivities to either visible wavelengths or to gustatory stimuli. When a chemosensory or visual mixture is presented, a new neural pattern occurs. Panel 2 represents a perceptual condition that differs from both analysis and synthesis. In this instance, simultaneous presentation of pure tones at 348, 440, and 524 Hz yields the perception of an auditory pattern, a musical chord (F major) that had not been perceived when any one or two of the three tones was presented. Nonetheless, some observers can still perceive and correctly identify the individual component sounds of this acoustic triad. Condition 2 can be taken as a pictorial example of fusion, although this was not specified by Erickson (1977). The absence of any change in illustrated neuronal activity between the "individual" and "together" neuron levels of panel 2 is, it is believed, logically inconsistent with the perceptual differences of the percepts level. Reproduced from Erickson, R. P. (1977), The role of 'primaries' in taste research, in J. Le Magnen and P. MacLeod (Eds.), *Olfaction and taste VI* (p. 371), by permission of Oxford University Press.

tures could also be perceived in three different manners. One would be analytically, composed of elements that could be identified; a second, synthetically, being solely taken as a new, unique percept. The third would provide coexistence of both conditions, that is, a novel percept for which the elements are still sometimes identifiable.

This alternative view submits that the proposed analytic or synthetic relationships between taste and the nature of gustatory perceptual processing neither exhaustively describe the logical possibilities nor adequately fit the empirical data. I suggest they are not suitable for much of taste perception (see Section III,B,4,a), and have little relation to flavor. As indicated, a third possibility can be considered. It has been named *fusion*. "In a fusion, the components of the fused sensation are still perceptible by careful analysis" [by an expert observer] (McBurney, 1986). The argument is that although a mixture of taste stimuli may indeed produce a new perceptual quality or entity that is absent from any one (or several) of the unmixed constituents, recognition of each of the individual qualities or stimuli will nonetheless be possible for sufficiently careful and practiced judges. Emergence of a new taste percept from a mixture would violate the analytical rule, while the possibility of recognition of each component is incompatible with the concept of synthesis.

Gustatory fusion is an interesting bridge between, or perhaps a substitute for, the analytic and the synthetic camps. Its roots are in 19th century philosophical psychology and the experimental study of mental laws of association (Boring, 1950). In that context, phenomenological observations by highly trained and experienced judges often found that fusion characterized, for example, complexes of musical tones (Boring, 1942). In the auditory domain, perceptual fusion states that a single tonal pattern or other multicomponent sound configuration can evoke a percept that is not only unique and coherent but is also analyzable by some observers of some or all of its components.

Similar outcomes have been reported for several studies of taste mixtures (Erickson & Covey, 1980; O'Mahony et al., 1983; Schiffman & Erickson, 1980, 1993). Although the findings of these gustatory experiments had considerably similarity, their experimental designs, and underlying rationales, were quite different. O'Mahony et al. (1983) had a strong methodological focus, whereas Erickson and colleagues emphasized a distinction between theories of taste perception.

The design of the Erickson and Covey (1980) studies, and more importantly their interpretation, have been questioned by Bartoshuk and Gent (1985) and Bartoshuk (1988). The objections are based on the proposal that in the two-component gustatory mixtures used by Erickson and Covey (1980), the taste of one component was necessarily suppressed by the other. The proposed suppression (see Lawless, 1979; Kroeze & Bartoshuk, 1985) would then explain why these mixtures were classified by subjects as having a "singular" taste rather than "more than one" taste. The Erickson and Covey (1980) experiments, their criticism by Bartoshuk and Gent (1985), and subsequent studies by Erickson and colleagues (1990) will be discussed in more detail in Section III,B,4,a. In that discussion, data relevant to gustatory fusion will be encountered.

A process that is revealed only through careful analysis by expert observers might seem to have little relevance to every-day taste-based decisions such as those outlined in Section I of this chapter. Nonetheless, it can be argued, and probably has been in sources that I have either failed to read or have forgotten, that the ability of some humans to report correctly the components of certain taste mixtures not only demonstrates that human taste processing can produce this outcome (assuming controls for separate learning of ingredients, as noted in the previous section) but also suggests that most humans probably have, under appropriate conditions, the ability to veridically dissect the new percept produced by a taste mixture into its constituents. Recall that Erickson and Covey (1980) did find that taste mixtures ranged from those that were judged to be almost completely "singular" to those that were predominately assigned to the other available category, "more-than-one." A fusion theory of gustation, in which novel taste percepts arise from mixtures but the mixtures can be perceptually decomposed (McBurney, 1986), may be closer to empirical reality at the central nervous system (CNS) processing and perceptual levels than either an analytical basic tastes model (Faurion, 1987b) or a synthetic pattern theory of taste. The explicit prediction for gustatory fusion would be that a single taste stimulus array can evoke a percept that is not only a distinct and original entity but is also analyzable by suitable observers with respect to some or all of its ingredients. Thus, in a fusion, the elements will be grouped into a unified perceptual object for all observers, and may be unitary or singular (as contrasted to more-than-one) for some or most observers, but at least a few observers will be capable of perceiving the components.

3. The Basic Tastes Theory

Basic tastes models seek to account for taste-dependent behavior by positing a very limited array of unique taste receptor-level processes, usually four, and correlated perceptions: sour, sweet, bitter, and salty (Bartoshuk, 1988; Beauchamp, Cowart, & Schmidt, 1991; Ganzevles & Kroeze, 1987; Kinnamon & Getchell, 1991; McBurney, 1974; Scott & Plata-Salaman, 1991). These four categories are considered to relate directly to chemically defined stimulus groupings: "Four corresponding compounds are recognized as eliciting these specific responses" (Zapsalis & Beck, 1985). The strict version of the basic tastes/taste primaries theory holds that quite exact connections exist between specific chemical stimuli, mutually exclusive response categories or types of taste receptors, and highly delimited, fully modular taste perception classes (Brand, Teeter, Kumazawa, Taufiqul, & Bayley, 1991; Kinnamon & Getchell, 1991; Scott & Plata-Salaman, 1991; Yamaguchi, 1987, 1991). Somewhat weaker or more qualified relationships are proposed by some workers in the food science community: "The respective taste primaries are associated with limited parallels to ions or molecular structures" (Zapsalis & Beck, 1985), or "As a general rule (with exceptions easy to find) sugars taste sweet, alkalide halides taste salty, acids taste sour, and alkaloids taste bitter" (Burgard & Kuznicki, 1990). A similar relationship had been suggested by McBurney (1984). Strong and rigid connections between the perceptual taste primaries and particular gustatory chemical stimuli are believed to permit definition of each taste receptor type and the particular perceptual process that the activity of a taste receptor type uniquely elicits in terms of chemically specified "type" stimuli (Scott & Plata-Salaman, 1991).

Based on these theoretical considerations, schemes for discovering taste coding are then followed in which the type chemical for a particular taste perception/ receptor category is the crucial probe. The ideal outcome for a test chemical is considered to be a behavioral response that is completely specified by, or totally unrelated to, one of the type chemicals (Kurihara, 1987; Scott & Plata-Salaman, 1991). Each such independent category is taken to be a *basic taste,* somewhat comparable to *best frequencies* in studies of auditory and other mechanoreceptors (see Barlow & Mollon, 1982; Pickles, 1988).

a. Tests of the Basic Taste Theory

If a sufficiently limited and carefully selected set of "simple" (Halpern, 1973) gustatory test stimuli is used, groupings that conform to the basic tastes model can be arranged. One has then identified the set of basic or primary taste stimuli and, of course, the basic tastes (Nishimura & Kato, 1988). However, a more representative assemblage of food-related or chemically diverse stimuli produces many theoretically inconvenient, intermediate outcomes (Faurion, 1987a, 1987b, 1994; Schiffman & Gill, 1987). Indeed, even the staunchest adherents of the four basic tastes model suggest that sweetness and bitterness, although two of the basic tastes, are not unitary classes because they may be separated into several and perhaps many subcategories using conventional taste psychophysics procedures (see Bartoshuk, 1987, 1988; Boughter & Whitney, 1993; Faurion, 1993; Faurion, Saito, & MacLeod, 1980; Rouseff, 1990; Schiffman et al., 1986; Schiffman & Erickson, 1993; Yokomukai, Cowart, & Beauchamp, 1993). Nonetheless, these subcategories of sweetness and bitterness are not considered by basic tastes theorists to be building blocks that combine to form a single basic taste perceptual category, but rather are considered each to somehow elicit fully the perceptual category (Miller & Bartoshuk, 1991). This may be a premature conclusion, because it is not clear that total or even substantial equivalence of the perceptual experiences evoked by the many sweet or bitter stimuli has been demonstrated.

Sour taste also seems to have a fundamental problem with bitter, another generally accepted basic or primary taste. The problem is that humans may not consistently distinguish between bitter and sour descriptions. This is referred to as sour–bitter "confusion" by those who believe that sour and bitter are independent, nonoverlapping, modular perceptual categories. Explanations based on shared receptor mechanisms are often provided. These explanations are neither useful nor convincing, because overlapping mechanism may be the rule rather than the exception in taste transduction and coding (Simon & Roper, 1993). Another analysis of sour–bitter confusion points out that subjects who initially use both sour and bitter to describe the same taste stimulus or several stimuli can be successfully instructed to use only one of these words for that stimulus, and the other for a different stimulus. This outcome is then interpreted as demonstrating a confusion that was eliminated by the instructions. However, this is far too strong a conclusion. A more conservative and parsimonious explanation is that the two solutions are discriminable, and that just as subjects were taught to call one bitter and the other sour, they could have learned to label one as Linda and the other as Frank, if that had been what the experimenter requested (see O'Mahony, Goldenberg, Stedmon, & Alford, 1979).

Other possibilities remain. One is that the relative activity of two (or more) mechanisms determines the probability of bitter versus sour taste quality descriptions. If this is correct, then the independent, mutually exclusive rule of the basic tastes model is violated, and the alternative pattern theory of taste perception (see the next section) is evoked.

b. Taste Matching with Mixtures of Primary Taste Stimuli

Empirical tests should allow evaluation of perceptual models. One application of the (four) basic tastes/taste primaries approach proposes a rough parallel to color matching in vision. It is argued that because the set of basic taste stimuli consists of four chemicals, A, B, C, and D, an appropriate mixture of all four will duplicate all possible tastes (Schiffman & Gill, 1987; Yamaguchi, 1991). The assertion that a set of four [or according to some, five (e.g., Nishimura & Kato, 1988)] "type" taste chemicals, mixed in an appropriate ratio, can duplicate any taste seems logically unavoidable if the underlying assumptions outlined above for the basic tastes model are valid. It has been asserted that successful results have been achieved (Van Skramlik, 1922, cited in Bartoshuk, 1980, 1988; Von Skramlik, 1926, cited in Pfaffmann, Bartoshuk, & McBurney, 1971 and in Beets, 1978). Erickson and Covey (1980) reported that mixtures of sucrose and fructose were judged to be 80% "singular" (as contrasted with "more-than-one"), which was close to the percentages obtained using solutions of sucrose or fructose alone. Recent experiments have found that a perfect match, that is, absence of discrimination, occurs for appropriate concentrations of the sugars glucose and fructose (Breslin, Kemp, & Beauchamp, 1994), even after equal amounts of NaCl are added to both solutions (Breslin, Beauchamp, & Pugh, 1996), as well as for the highly dissociated inorganic acids HCl, H_2SO_4, and HNO_3 when at equal pH (Breslin & Beauchamp, 1994).

Nevertheless, failure is the regular outcome of less highly restricted psychophysical tests of this theory-based matching taste experiment (e.g., Konosu, 1979; Konosu, Yamaguchi, & Hayashi, 1987; Solms & Wyler, 1979). One might suppose that this would doom the basic tastes/taste primaries models. However, as has often been pointed out, mere facts and empirical observations do not destroy theories. In this instance, two routes are generally taken to sustain the model. The more common path is to simply ignore the results, or to suggest that they are not germane because they use food or food-derived chemicals as the stimulus for the taste to be matched, rather than a more appropriate simple stimulus. The second approach is to explain that the apparent taste is really not a taste at all. Instead, it is dependent, at least in part, on chemicals that are not proper taste stimuli. The so-called alkaline, metallic, and meaty tastes (Zapsalis and Beck, 1985), the perceived oral drying, puckering, and roughing (the astringency complex) (Lawless, Corrigan, & Lee, 1994) of the tongue due to substances such as polyphenols (tannins) (Delcour, Vandenberghe, Corten, & Dondeyne, 1984; Noble, 1990), and the chemically produced "heat" from capsaicins and similar molecules (Green & Lawless, 1991; Silver & Finger, 1991) are all cast into this "not taste" pit by the basic taste theorists (Hettinger & Frank, 1992; Lawless, 1991; Snow et al., 1991). These exclusions are rarely justified by empirical data [although Breslin et al. (1993) claim to have excluded astringency from the taste domain, and Hettinger, Myers, & Frank (1990) have shown that nasal chemoreception can be necessary for some sulfurous, soapy, or metallic reports, although not with monosodium glutamate]. It appears that a wish to support a particular theoretical position, or a tendency to follow the categorizations used by previous workers in a rather unquestioning manner, are common biases.

c. Receptor and Neural Applications of the Basic Tastes Model

Physiological investigations based on the independent-channels, basic tastes model identify those sensory elements that have the lowest threshold to the type chemical or give the largest response of some sort to it. For type-stimulus A, these elements are then designated A-best. It is usually assumed that responses to all other chemicals, which of course do occur, have no sensory meaning and consequently are not constituents of a "true" chemosensory code (Uttal, 1969). This strategy is rather similar to those auditory or somesthetic studies that focus on frequency-related tuning curves and bandwidth measures, and determination of the frequency at which the least energy is needed to elicit a criterion response (the best or characteristic frequency). Single-frequency sinusoids are used as stimuli, and define all aspects of the sensory system. Naturally occurring sounds or vibrations, which are almost always complex and modulated, are not easily accommodated by the models derived from this pattern (Capranica, 1992). A comparable difficulty, in this case of applicability to foods and beverages, devolves to taste theories that require specificity and independence of component processes from receptors through perception. It is of considerable importance that some current analyses of taste physiology take a much more flexible and eclectic approach, combining moderately but not totally selective taste receptor types with the necessity of across-type patterns for adequate central nervous system processing (e.g., Smith & Frank, 1993).

4. Pattern Models of Taste Perception

The multidimensional and diverse nature of the chemicals that are effective taste stimuli is both an inducement and a problem for the pattern theorists (Schiffman & Gill, 1987). For this approach, a numerically limited (but generally exceeding four) set of receptor-level elements with *overlapping sensitivities* is assumed (Erickson, 1984; Faurion, 1987a, 1987b). The underlying model is color vision, with the pattern of *relative response* across a limited array of receptor types able to encode uniquely a wide range of stimulus events and, after necessary processing, evoke many perceptual events (Erickson, 1963; Erickson & Covey, 1980; Schiffman & Erickson, 1980; 1993; Faurion, 1994). An immediate problem is specification of behavioral operations with taste stimuli that will verify or falsify these concepts of pattern-based taste perception.

a. Responses to Mixtures: A Crucial Test of the Pattern Theory

An important approach to the need for tests of the pattern model of taste perception involves the use of mixtures. The reasoning is that if the way in which something tastes, that is, taste quality perception, is due to the relative activity of several interrelated and overlapping processes, then descriptions of some mixtures will be different from the perceptions evoked by any one of the component chemicals (Erickson, 1985; Erickson & Covey, 1980; Erickson et al., 1990). In addition, it is sometimes stated, in analogy with the results of color mixtures, that the component chemicals of a taste mixture will not be identifiable: "psychophysical data suggest that separate tastes, such as the primary four, may not remain separate, but may combine in ways in which the original components are lost, and thus tastes other than the primary four must be produced" (Erickson, 1985). Conversely, proponents of the basic tastes model often argue that a mixture must be nothing more than its elements, with the individual components recognizable as long as the concentrations are of similar but not saturating effective intensities: "When substances are mixed, the taste qualities of the components are identifiable but the perceptions of the intensities of those qualities are often suppressed" and "Taste mixtures are very different from color mixtures. . . . When taste qualities are mixed, the qualities of the components can be identified" (Bartoshuk, 1988).

   A number of experiments have examined psychophysical responses to taste mixtures in relation to theories of taste perception or analytic versus synthetic models for gustatory perception (e.g., O'Mahony & Buteau, 1982; O'Mahony et al., 1983; Schiffman & Erickson, 1980, 1993). Erickson and Covey (1980) proposed that "The issue of whether or not mixtures of different tastes are perceived as "unitary" provides a test of the four-primary model" and that the synthetic model of taste perception requires that "single taste sensations—such as the primaries—should fuse, when combined, to form singular sensations." In order to demonstrate that their psychophysical procedure, in which every stimulus was to be classified as either "more-than-one" or "singular," was a suitable test for both analytical and synthetic perceptual modes, the experimenters had subjects categorize three different arrays of visual or auditory stimuli: calibrated colored papers (15 Munsell colors, ranging from the red 5R 5/14 to the purple 5P 4/8) (see Wyszecki, 1986), overlapped monochromatic lights, and recorded piano notes presented either singly or paired and separated by a half-step (a semitone: "an interval equal to a half tone in the standard diatonic scale") to an octave (the interval of eight diatonic degrees between two tones, one of which has twice as many vibrations per second as the other) (American Heritage Electronic Dictionary, 1993). Erickson and Covey (1980) found that the colored papers, the monochromatic light mixtures, and the individual piano notes were "singular" and the pairs of piano notes were "more-than-one." These outcomes were as anticipated for typically synthetic color perception and typically analytic perception of sufficiently separated tonal pairs.

Their gustatory experiment (Erickson & Covey, 1980) asked subjects to classify the taste of solutions as either singular or more than one. As might be expected, mixtures of the sugars sucrose and fructose were about as singular as were solutions containing only sucrose or fructose: more than 80% of the trials for the mixture and for each saccharide alone were singular. However, comparable results were also obtained for mixtures of HCl and NaCl (often specified as the archetypical stimuli for two different and distinct basic or primary tastes, sour and salty), of HNO_3 and NaCl (commonly considered to be sour and salty primary stimuli), and of quinine sulfate and HCl (frequently prototypical stimuli for the primary tastes bitter and sour). Perhaps surprisingly, some mixtures were judged to be singular more often than one of the components. For example, the mixture of quinine sulfate and HCl was judged singular on 85% of trials, whereas quinine sulfate was judged singular on only 73% of the trials; the mixture of HNO_3 and NaCl, on 75% of trials, but HNO_3 on 63%. An extreme instance may be sodium acetate, which was judged singular on slightly more than one-third of the trials (38%) when tasted alone, but resulted in two-component mixtures that were rated singular on about two-thirds of the trials (when mixed with quinine sulfate or with KCl). On the other hand, mixtures of sucrose or fructose with the same metallic and organic salts, inorganic acids, or the alkaloid quinine sulfate gave a low incidence of rating as singular: a median of only 30% (range 11 to 64%). These two sugars, tasted alone, had been judged singular on 85% of the trials.

Erickson and Covey (1980) summarized the results in the following statement:

> The most conspicuous aspect of the gustatory data are that single stimuli, including the "primary" stimuli, are not consistently judged as singular, and that mixtures are not consistently perceived as being more than one. . . . It may be that the so-called primary stimuli are those single stimuli that are somewhat more singular than others, even though not forming a class distinct from them.

They also pointed out that

> There are problems here for both the synthetic . . . and analytic . . . points of view. According to a strict synthetic point of view, all stimuli and all stimulus mixtures should be rated as singular. . . . The problem for the analytic position is also clear . . . the mixtures . . . should all appear as more than one (except perhaps for the sucrose–fructose mixture): the synthetic mixtures of several "primary" stimuli are particularly problematic in this regard.

The Erickson and Covey (1980) final conclusion was that their data, in particular the high incidence of singular judgments for mixtures of what are often considered to be primary taste stimuli, were not compatible with the concept that four categories or tastes constitute the complete set of taste perceptions. Thus they argued that both the four-tastes version of the basic tastes theory and an exclusively analytical model of taste perception were rejected by the results of their experiment.

These data were interpreted very differently by Bartoshuk and Gent (1985) and Bartoshuk (1988). Gustatory mixture suppression, in which the perceived intensity of one or both components of a two-component mixture is reduced to a lower, and sometimes negligible, level (Lawless, 1979), was offered by Bartoshuk and Gent (1985) as the explanation for the high incidence of "singular" judgments found by Erickson and Covey (1980) for such mixtures:

> When two substances are mixed, even if their intensities are prematched (and those of stimuli used in the study of Erickson and Covey were not), one component is often more suppressed than the other. Thus in many mixtures only one taste quality is perceived.

Bartoshuk (1988), referring to the Erickson and Covey (1980) interpretation of their taste mixture classification data, stated:

> Unfortunately, this analysis ignored the large literature on mixture suppression. If one mixture component is greater than the other or has greater suppressive power, then the other component may disappear in the mixture. Needless to say, such a mixture is "unitary."

These reinterpretations of the meaning of the Erickson and Covey (1980) taste mixture observations constitute a testable hypothesis. Erickson et al. (1990), who designed an experiment to test this postulate, expressed it in the following manner:

> If a mixture of two singular tastes is perceived as singular because the "suppressor" stimulus erased the sensation from the "suppressed" stimulus, then the mixture would indeed be singular but it also should taste the same as the "suppressor" stimulus.

The Erickson et al. (1990) subjects were presented with groups of three solutions, composed of two identical solutions and one different solution. The task was always to select the one solution that was different from the other two [this procedure is called triangle testing (see Frijters, Blauw, & Vermaat, 1982; O'Mahony, 1986; Meilgaard et al., 1991)]. Subjects were also allowed, but not required, to indicate that one or more of the samples differed in intensity from the others of a triad. Mixtures of NaCl with HCl, $MgCl_2$, or NH_4Cl, and of HCl with quinine sulfate, as well as solutions of the components, were used. All had previously been employed by Erickson and Covey (1980). For the Erickson et al. (1990) study, concentrations of each component, by itself, that were comparable to the intensity of each mixture, were determined. Groups of three solutions consisted of two samples of the same mixture and one component of that mixture matched for intensity to the mixture, or one sample of a mixture and two identical samples of one component of that mixture matched for intensity to the mixture. Subjects either tasted each solution, or separately, smelled the solutions without tasting them.

Erickson et al. (1990) found that the mixtures of NaCl with HCl or $MgCl_2$, and of HCl with quinine sulfate, were consistently discriminated from each of their components when they were tasted ($p < .0001$), but this occurred for none of the mixtures when they were only smelled. Reports of differences in intensity were at a chance level for both smelling and tasting these mixtures and their components. Gustatory discrimination of the NaCl and NH_4Cl mixture from NH_4Cl was also highly significant, but was only just significant ($p = .044$) for NaCl. Again, olfactory discriminations were not significant, nor were reports of intensity differences.

It will come as no surprise that Erickson et al. (1990) concluded: "Since the taste of these mixtures must be other than the primary tastes used to form them, these findings question the presumption that only four 'primaries' are adequate to describe the range of tastes." This seems a rather mild and circumspect summary statement. Nonetheless, it avers that the discriminations between the mixtures and their intensity-matched components reported by Erickson et al. (1990) were based only on differences in taste quality, that is, differences in what the mixtures and the components tasted like, and not on disparities in taste intensity or nongustatory factors.

Is this claim justified? The nongustatory factors aspect is surely not controversial. Discrimination based on olfaction was explicitly tested; no evidence for such discrimination was found.

Intensity is a more complex question. Subjects were not required to report whether one or more solutions of each triad exceeded others in intensity, although they were permitted to do so. This could present a serious problem, because the hypothesis that the Erickson et al. (1990) experiment had been designed to test, offered by Bartoshuk and Gent (1985) and Bartoshuk (1988), proposed that large intensity changes due to mixture suppression were the explanation for the "singular" classification found by Erickson and Covey (1980) for many mixtures.

Erickson et al. (1990) found that on 82% of the trials subjects reported intensity differences between the three liquids presented. However, subjects were equally likely to state that the mixture or a solution of a component was more intense. Consequently, separate statistical evaluations of intensity categorizations for each mixture and its components showed no significant differences ($p \geq .08$, median $p = .24$). Evidently, the intensity matching between single-component solutions and mixtures that Erickson et al. (1990) did in their pilot study was successful. An absence of consistent intensity differences eliminates the mixture suppression explanation proposed by Bartoshuk and Gent (1985) and Bartoshuk (1988) as the mechanism underlying the "singular" perceptual classification of two-component taste stimulus mixtures reported by Erickson and Covey (1980). This is the case because the suggested mixture suppression was hypothesized to cause so much attenuation of the taste intensity of one component that a singular rather than more-than-one percept resulted.

With taste intensity differences leading to mixture suppression thus excluded as bases for the observed discriminations between the binary mixtures and their components, and olfaction having already been barred as an explanation, it appears reasonable to conclude that the taste of each mixture did not correspond to the taste of either of its components. Does this outcome validate the pattern theory of taste perception and falsify the basic tastes theory? Could the number of basic tastes simply be expanded? Because the components of the solutions used by Erickson et al. (1990) included three of the four standard stimuli for basic tastes, perhaps Erickson et al. (1990) actually demonstrated three new basic tastes. If we add sucrose, which was not used, do we now have eight basic tastes? Would more such experiments give 20, 50, or 100 basic tastes? At some point, we would then abandon the concept of basic or primary tastes, just as we do not find a concept of basic or primary sounds or tones useful.

Although the just-depicted expansion *ad absurdum* of putative gustatory primaries has not been reported, it might well occur if mixture experiments similar to those of Erickson et al. (1990) were done with sufficiently large and diverse arrays of taste stimuli. This outcome would be sufficient to eliminate the basic tastes theory as presently constituted, but would not necessarily support a pattern theory of taste requiring that synthesis be the only or the major perceptual process. Instead, each of the hypothetical many discriminable tastes could be analogous to the numerous discriminable tones of auditory perception.

At this point one might object that the multitude of separate tastes under discussion would be revealed only when taste mixtures were produced; that is, they appear to depend on a synthetic process. However, I suggest that if the discrimination test used by Erickson et al. (1990) was applied directly to comparisons of the tastes of the many diverse molecules that are chemical taste stimuli (see Section III,A,2,a), myriad distinguishable tastes would be illustrated in the absence of mixtures. Perhaps ability to discriminate tastes, although quite appropriate to demonstrate that mixture suppression was not the explanation for the Erickson and Covey (1980) observation that taste mixtures were often perceived as singular, cannot be used to clarify other issues of gustatory categories or groupings. The singular versus more-than-one dichotomy may also be problematic. These considerations are discussed in the next section.

b. Fuzzy Reality and All-or-Nothing Concepts

Gustatory and other research hypotheses are commonly couched as mutually exclusive outcomes, thus permitting an "either-or" evaluation that allows a sequential "if-then" path of theory and test to be followed. This design is a useful and perhaps necessary practice, meshing well with statistical aspects of null hypothesis testing. The empirical measures selected to falsify or verify these hypotheses are sometimes designed to facilitate categorical evaluations, thus corresponding to the mutually exclusive character of the underlying hypotheses and of typical inferential statistics.

i. "Singular" versus "more-than-one" With reference to the processes by which perception of mixtures occurs, the singular versus more-than-one classification employed by Erickson and Covey (1980) implies that taste and other experiences can only fit one of these two categories. Pairs of tones or colored lights were indeed shown to correspond almost exclusively to the more-than-one and the singular categories, respectively. However, as discussed earlier (Section III,B,2) tonal and other sound patterns can represent a unique, essentially singular, percept. Nonetheless, several of the components of this percept can sometimes be identified, thus earning a more-than-one category. This combination of putatively incompatible characteristics illustrates the hazard of requiring that all mixture perception correspond to one of two quite delimited classes.

ii. Discrimination   Discrimination tests represent another seemingly simple but sometimes rather complex situation. If the crucial question is the presence or absence of discrimination, per se, as between mixtures and component solutions in Erickson et al. (1990), there is no particular problem. However, use of a discrimination measure to establish taste classifications or categories may be much less straightforward. When taste stimuli are discriminable, what has been established beyond the fact that they can be consistently distinguished? Can we assert that discriminable tastes are totally independent, with no common attributes? The literature on categories and categorization informs us that groupings and distinctions can be made at a variety of levels (Rosch, Mervis, Gray, Johnson, & Boyes-Braem, 1976; Medin, 1989). With reference to taste, descriptions of single-component solutions made from a variety of molecules are likely to include the term *sweet,* yet many of these solutions are discriminable from each other (Bartoshuk, 1988; Faurion et al., 1980). A comparable situation exists for *bitter* (Boughter & Whitney, 1993; Breslin, 1996; Rouseff, 1990). It is clear that some discriminable tastes will nonetheless be placed into a common category by many observers. Tastes, as well as chairs, trees, birds, and so on, can be discriminable and also, in some instances, more or less similar (Harnad, 1987; Medin, 1989).

iii. Degree of similarity   If certain taste stimuli are considered to be very similar to each other, but less similar to all other taste stimuli, each collection of related stimuli might define a perceptually (and therefore, physiologically) meaningful stimulus array. This apparently simple proposal has been adopted by many investigators. Measuring the degree of similarity has presented difficulties. Far more problematic have been attempts to define an overall relationship between taste stimuli by using similarity data to specify degree of hierarchical clustering or a multidimensional taste stimulus space. The basic difficulty is that hierarchical clusters, and orderings in *n*-dimensional space, will appear from random number input (Erickson, Rodgers, & Sarle, 1993). Consequently, it seems that hierarchical cluster analysis and multidimensional scaling, although useful for experimental data analyses that will generate hypotheses, may not be able to verify or falsify hypotheses.

C. Is Taste an Ongoing Process That Should Be Studied over Time?

1. In the Beginning

In 1937 a study of human taste perception over 10-sec time periods was published (Holway and Hurvich, 1937). Each subject had been instructed to draw a graph of taste intensity over time, starting from the moment at which a drop of solution was placed on their extended (through closed lips) tongue. The principal findings were that higher stimulus concentrations evoked time–intensity functions with steeper initial slopes, and that maximum intensity was greater for higher concentrations but was reached later. From these observations, the authors concluded that

> Ordinarily the concentration of the gustatory substances is regarded as the stimulus to intensity. . . . However, . . . while the concentration originally placed on the tongue is 'fixed,' the intensity varies in a definite manner from moment to moment. . . . Intensity depends upon time as well as concentration.

This early description of human taste perception as a dynamic sequence in which the intensity of tastes increases during a time course of seconds has been confirmed and extended by numerous laboratory investigations (reviewed in Halpern, 1991). The procedure is well known and currently used in the food science community (see Ayya & Lawless, 1992; Bonnans & Noble, 1993; Lawless & Clark, 1992; Robichaud & Noble, 1990; van Buuren, 1992). That taste is an ongoing process that changes over time may not be a surprise of historic proportions to most readers. The major requirement for knowing that time is an important factor in taste is the habit of oral ingestion of foods or beverages (see Köster, 1981). Anyone who has taken food or drink by mouth, and has given any attention while doing so, is already aware that tastes develop, peak, and change during and between the sips, bites, and swallows of normal eating and drinking.

2. Perceived Taste Changes; Other Perceptions Do Not?

The observation that perceived tastes are commonly expected to change both during and immediately after single sips of a beverage or bites of food may set taste apart from other perceptual systems. If the size, shape, or color of the text you are reading appeared to you to change, you might suspect some defect in your own perceptual systems, or at least some alteration of the light in the room or properties of the display. If you were using a sound recording or an audio communications system that changed its range of pitch, or began waxing and waning, you would suspect that the system, or possibly your hearing, was faulty. Why this difference between taste and hearing or vision?

Paradoxically, it may be because oral chemosensory stimuli tend to be much more constant than sights and sounds. We frequently move our eyes, head, and body, producing rapid changes in auditory and visual input. Sounds themselves are never constant. Those we hear best change intensity thousands of times per second. Communicative sounds such as speech have complex temporal patterns. Thus, hearing and vision deal with stimuli that are always changing rapidly in certain ways. From this array of changing acoustic and optical events we usually perceive at least short-term constancy.

Taste stimuli also change as we eat, drink, chew, or swallow. However, the time scale is much more leisurely than that of sounds or lights. Our sips have durations of about 1 sec, and we will retain liquids in the mouth from one-half to several seconds before swallowing. Our chewing has a similar timing—once or twice per second. Thus taste stimuli may remain relatively constant for periods from one-half up to several seconds (Halpern, 1985, 1991; Delconte et al., 1992). When they change it is largely under our control, within our mouth. Still, the foregoing claim may oversimplify the situation. A proper evaluation would require that one know the importance for taste perception of interactions between taste receptor organs and taste stimuli as well as changes in the stimuli during normal ingestion. Factors such as salivary dilution, tongue movement, and effects due to chewing would be relevant.

Eating and drinking produce a sequence of changes in the mouth, as noted above, but perceived changes in taste can also occur while the physical conditions at the taste receptors remain constant. Decreases in taste intensity or a diminution and disappearance of taste quality during prolonged constant conditions at taste receptors are not surprises, because under unchanging circumstances adaptation is commonly encountered in sensory systems (e.g., Levine & Shefner, 1991; Sekuler & Blake, 1994). In contrast, during the first few seconds of constant taste stimulation, taste intensity is found to increase, while taste quality descriptions appear and then disappear (Halpern, 1991, 1994; Zwillinger & Halpern, 1991). If gustatory stimulus duration exceeds about 2 sec, judged intensity often approaches an asymptote. Finally, when taste stimulation of 4 sec or less ends, perceived intensity declines much more slowly than the time course of stimulus molecule removal from the tongue. The rationale for limiting taste stimulus durations to 4 sec when studying taste perception over time, the methods used, and a summary of the results are given in the following discussions. Other approaches for examining the time course of taste perception, which represent the majority of all such research, are reviewed in Halpern (1991). More recent examples can be found in reports by Ayya and Lawless (1992), Bonnans and Noble (1993), Robichaud and Noble (1990), and van Buuren (1992).

3. Time—Intensity and Time—Quality Studies of Taste

a. Appropriate Stimulus Durations

Specification of the time periods during which taste stimuli are present and over which taste perception is to be measured has an immediate connection to examinations of the temporal nature of taste. Direct laboratory investigations of human drinking behavior indicated that the time interval during which liquids were held in the mouth before being swallowed, or perhaps spit out if that was an option, depended on the task presented to the subjects. If subjects were asked only to take one sip from a glass and then always swallow the liquid, with no judgments requested or overt taste-dependent decisions to be made, contact duration with the liquid in the glass was about 1 sec, with swallowing in the next 0.5 to 0.75 sec (Halpern, 1985). This pattern occurred for H_2O, for Kool-Aid, and for Kool-Aid with NaCl added. Thus, for a single sip in relation to which no decision or report was required, taste stimulus duration from the beginning of contact with the liquid until swallowing was about 1.5 sec, although total effective stimulus duration might have been longer due to stimuli remaining in saliva or sorbed onto the tongue (Halpern, 1986b).

Quite different human drinking behaviors occurred when a taste-dependent decision to swallow or spit out was required (Delconte et al., 1992). The task was to taste a liquid, and then, based on that taste (the target taste), with no linguistic description made available to the subject or requested from the subject, to *swallow* a single sip from each of a series of sample glasses if the taste did *not* correspond to the target taste but to *spit out* the sip if the taste *did* correspond to the target taste. Under those circumstances, times from contact with a liquid to swallow were about 2.5 to 3.5 sec when a correct decision to swallow was made (the majority of all swallows), but about 4 to 5.5 sec when incorrect swallows were made (i.e., the sipped sample liquid was the same solution as the originally tasted target liquid, and should have been spit out). Time intervals from contact to correct spits (the majority of all spits) ranged from about 3 to 4 sec. Incorrect spits (i.e., the sipped sample liquid was not the same solution as the tasted target solution, and should have been swallowed), which were quite uncommon for most liquids, could have durations as brief as 1.5 sec from contact until spitting out.

These two sets of experiments indicate that human drinking behavior typically produces taste stimulus durations that range from a minimum of 1 to 2 sec up to 4 to 5 sec at the upper end. Briefer gustatory stimulus durations, although fully competent under laboratory conditions to elicit human judgments of taste intensity and quality (Kelling and Halpern, 1988) and to permit nonlinguistic taste intensity tracking over time (Halpern, 1991), are probably below the limit of normal human experience. For the long-duration aspect, 4 or at the most 5 sec of continuous stimulation may represent the upper temporal boundary of familiar human taste events.

b. Methods for High Temporal Resolution Time–Perception Tracking

In order to study perceptual responses to gustatory stimulus durations of 1 to 4 sec, measurements must be made many times each second during the period of time over which responses occur. In contrast, most time–intensity studies have measured taste responses at relatively low temporal resolutions over 30- to 180-sec stimulus durations (see Halpern, 1991; van Buuren, 1992).

i. Intensity   High temporal resolution measurements of taste intensity responses over time have been secured by obtaining and digitizing the tracking responses of subjects once every 100 msec, that is, 10 times per second for 8 to 10 sec after solution flow began (Halpern, 1991). The resultant 100-msec resolution is appreciably faster than human taste reaction time (see Halpern, 1986a; Kelling & Halpern, 1987). Subjects tracked taste intensity by using a single-axis joystick to control the vertical position of a computer display, which they viewed during highly controlled liquid flow over their tongue. Instructions were to track total taste intensity. The totally nonlinguistic intensity tracking just described may be necessary for the gustatory time–intensity data that will be discussed in Section III,C,3,c. This is suggested because our attempt to do language-based taste–intensity tracking using touch typing and the numeric keys on a standard computer keyboard produced results comparable to single-judgment gustatory magnitude estimations: one or two key presses about 1.5 to 2 sec after stimulus flow at the tongue began. Attempts to use a nonlinguistic key pad for taste intensity tracking also suggest that use of a 9-key pad as a positioning device may be quite different from taste–intensity tracking using a single-axis joystick (J. S. Meltzer, M. Y. Lee, & B. P. Halpern, unpublished observations, 1993).

ii. Quality descriptions   Time–quality tracking has been done using a learned, 23-item taste–quality descriptor list (Halpern, 1991; Zwillinger & Halpern, 1991). The descriptors were represented in a learned code by single letters. The task of each subject was to track the taste of the liquid during highly controlled liquid flow over her tongue (see Kelling and Halpern, 1986). Subjects made their responses with touch typing on the alphabetical keys of a standard computer keyboard, for which 2-msec resolution of response times was provided. Feedback of each key press was provided on a computer monitor, using updates every 100 msec.

Time–quality tracking stimulus liquids were 2 mM sodium saccharin, 214 mM monosodium glutamate, H_2O, and a mixture of 10 mM citric acid in 2 mM sodium saccharin, all at a 1-sec duration. Comparisons of time–quality tracking data with time–intensity tracking outcomes will be described later. However, the differences in response measures for the quality description and intensity tracking data sets should be noted, because they raise the possibility that any observed disparities in temporal patterns may be due to methodology rather than to dissimilar mechanisms of gustatory perception.

c. High Temporal Resolution Time–Perception Data

i. Time–intensity   For stimulus durations of 1 sec or longer, intensity tracking began within 1 sec of solution arrival at the tongue. Perceived intensity then increased rapidly (Figure 2), reaching 90 to 100% of maximum within the next second (Halpern, 1991, 1994). With 1-sec stimulus durations, tracked taste intensity reached maximum within a few hundred milliseconds after stimulus removal; with 2-sec stimulus durations, the maximum occurred within a few hundred milliseconds before or after the end of stimulus presentation (Halpern, 1991). For 4-sec stimulations of the tongue, median perceived taste magnitude was at ≥90% of maximum intensity throughout the latter part of stimulation, after the initial rapid climb in intensity during the first 2 sec.

In general, the taste intensity sequence observed with high-resolution tracking was a rapid increase in taste intensity that slowed after about the first second, continued for 1 or 2 sec if stimulation continued, and then remained at a more or less constant level for the next second of stimulation. Both higher stimulus concentrations and longer durations led to greater maximum intensities.

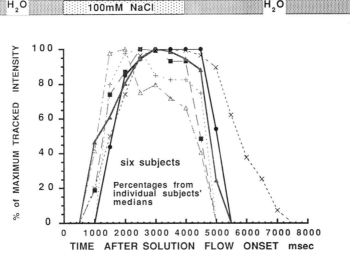

FIGURE 2   Time–intensity tracking of total taste intensity. Presentations for durations of 4 sec of an aqueous solution of 100 mM NaCl; the solutions flowed at 10 ml/sec over 39.3 mm[2] of the antero-dorsal tongue tip region of trained adult humans and were tracked during eight replications by each of six subjects (Halpern et al., 1992; Halpern, Meltzer, & Darlington, 1993). The data plotted in the graph are percentages for each subject, calculated by dividing the 80 median tracking intensities for each subject (100-msec temporal resolution of joystick position, 10/sec for 80 sec) by the maximum intensity value for that subject, and multiplying by 100. For details of stimulus presentation procedure, see Kelling and Halpern (1986); for high temporal resolution time–intensity tracking, see text and Halpern (1991).

When taste stimulation ended, tracked taste intensity decreased slowly, despite a rapid removal of stimulus solutions from the tongue (Halpern, 1986b; Kelling and Halpern, 1986). For stimulus durations of 4 sec, perceived taste duration regularly exceeded solution duration at the tongue by at least 1 sec, and sometimes by 3 to 4 sec (Halpern, 1994). This disparity of 1 sec or more between physical stimulus presence at the tongue and perceived intensity was also observed for 1-sec stimulus durations (Halpern, 1991).

In summary, perceived taste intensity changed dramatically during stimulus durations of 1 to 4 sec, which corresponds to the limits of intraoral stimulation during normal drinking and swallowing (or spitting out). Study of this dynamic process requires ongoing, high-resolution measures. Relationships between particular taste stimuli and characteristics of these time–intensity patterns remain to be resolved.

ii. Time–quality   Taste quality descriptions also changed over time. Onset of first key press for quality tracking of sodium saccharin ranged across subjects from 937 to 1353 msec after solution flow onset (Zwillinger & Halpern, 1991). That time interval was comparable to the vocal taste quality reaction times observed for sodium saccharin when a number of different taste stimuli were randomly presented (Halpern, 1987). The taste quality key presses for sodium saccharin indicated sweet on 75% of the trials and sugar on 6.5%. These responses were similar to verbal descriptions of sodium saccharin taste quality (Kelling & Halpern, 1988). However, in less than 1 sec after the key press for sweet or sugar, subjects pressed the key that indicated "no taste." For the 1-sec stimulus durations, total quality duration for 2 mM sodium saccharin was 594 ± 11 msec (median ± standard error of the median), with median quality durations for individual subjects ranging from 515 to 682 msec. Similar brief taste quality durations were observed for 214 mM monosodium glutamate and a mixture of 10 mM citric acid in 2 mM sodium saccharin (557 ± 23 msec and 657 ± 23 msec, respective median durations ± standard error). These brief taste quality durations were surprising because, as discussed in the previous section, tracked taste intensity durations were much longer than stimulus duration.

It would be intriguing if taste quality durations were truly shorter than perceived taste intensity (Halpern, 1991). Other time-related fundamental differences between taste intensity and taste quality judgments have been observed. For example, verbal taste quality descriptions for NaCl, sodium saccharin, HCl, and a mixture of 10 mM citric acid in 2 mM sodium saccharin showed little or no modification when taste stimulus durations increased from 50 to 2000 msec, but magnitude estimates of their taste intensity more than doubled over this range of durations (Kelling & Halpern, 1988). Nonetheless, it would be premature to conclude that taste quality durations are actually briefer than perceived taste intensity. The problem, as suggested above, is that linguistic keyboard input was used for taste quality tracking, whereas nonlinguistic responses with a joystick were used for the taste intensity tracking.

For monosodium glutamate, time–quality tracking revealed that two separate taste descriptors were used, separated in time, on 16% of the trials (Halpern, 1991). Previous verbal taste quality descriptions for monosodium glutamate had contained a number of different words, including two- and three-component adjective compounds. (Halpern, 1987; Kelling & Halpern, 1988). The most common of these verbal descriptors for monosodium glutamate had changed with stimulus duration, suggesting a temporal sensitivity that differed from other stimuli. Time–quality tracking data indicated that rather than a single very complex percept, monosodium glutamate evokes first one percept, then another, with a time interval of almost 1 sec between the taste quality percepts (Zwillinger & Halpern, 1991).

D. What Are the Proper or Permissible Responses to the Question: "What Does It Taste Like?"

1. General Considerations

For any perceptual domain, one can envision at least two approaches to a need for descriptions of the perceptual experiences evoked by environmental events. Perhaps the simplest and most obvious is to ask observers, "What does it (sound, smell, feel, look) taste like?" Cultural, social class, and lexical factors will of course have substantial effects on the responses. In general, however, observers will often answer with names of objects or processes with which they are familiar, for example, "it sounds like two cats fighting" or "it's the alarm call of the owl monkey." Under certain circumstances individuals will refer to an operational measurement scale that they and others use, for example, "the sugar solution is at the hard-ball stage" (Kander, 1954) or "it's as long as a football field." More rarely, observers will use an exact physical description, such as "it smells like amyl acetate," "the building is 10 meters high," or "the wavelength of the light is about 512 nanometers."

An alternative approach might be to provide observers with a set of descriptors, and instruct the observers to select all descriptors from this set. In effect, this may have already been done by conventional formal education for those who used exact physical descriptions when asked an open-ended question. Of course, the very same person, when asked what a musical passage sounds like, what a visual scene looks like, or how a massage feels, probably will not respond in terms of sound frequencies, specific musical notes, stated wavelengths of light, or force per square centimeter applied at a particular rate.

2. Taste Descriptions

a. Acceptable Words

When subjects are asked to indicate what something tastes like, how constrained or unconstrained should their responses be? One approach, as suggested previously, is simply to allow the subjects to say whatever they wish, and to accept whatever they say as the description. The rationale asserts that unrestricted descriptions will be the most unbiased and most valid characterization of taste stimuli, because they represent (it is hoped) the same linguistic taste responses used in everyday life. Free-choice profiling is a name applied to this method in the food science community (Rubico & McDaniel, 1992). Groups of subjects, doing free-choice profiling of acids with a "sip-and-spit" stimulus presentation technique, produced results that were comparable to those that had been obtained with highly trained taste panels.

When this approach was taken under very controlled laboratory conditions, most taste quality descriptions tended to be similar across subjects and to consist of a relatively small set of words (Halpern, 1987). In those experiments, liquids were flowed over the anterodorsal tongue using a procedure that both allowed liquids to contact only the desired region of the tongue and prevented all vapor phase (olfactory, vomeronasal, airborne trigeminal) stimulation (Kelling & Halpern, 1986). Under these circumstances, the majority of the descriptions for NaCl was "salty"; for sodium saccharin and for sucrose, "sweet"; for $MgSO_4$, "bitter"; and for tartaric acid, "sour" (Halpern, 1987; Kelling & Halpern, 1988), even as might have been predicted by the basic tastes theory. However, for other stimuli, such as HCl, monosodium glutamate, a mixture of citric acid and sodium saccharin, and a solution of a commercial beverage base, cherry Kool-Aid, to which sodium saccharin was added, no one descriptor represented a majority of the responses. Perhaps of more importance, adjective compounds, including salty–sour–sweet or sweet–lemon, as well as nonstandard (and, according to some, nontaste) terms such as metallic or soap, represented, in the aggregate, one-fifth of all responses to HCl, cherry Kool-Aid, or $MgSO_4$.

The relatively high incidence of adjective compounds and of words other than sweet, sour, salty, and bitter, when all descriptors are permitted, may indicate that limitation of acceptable taste responses to very few descriptors, each to be used in isolation, produces a rather distorted and excessively simplistic picture of human taste perception. In effect, allowing only the four descriptors specified by the basic tastes theory could yield a self-fulfilling prophecy. However, perhaps the complexity of the data is illusory. It could be that the adjective compounds or nonbasic taste descriptors are only synonyms for the basic taste terms, and do not represent any additional categories.

Direct study of taste category formation, a very useful approach to the question of taste perceptual categories, has been explored by O'Mahony and colleagues (see O'Mahony, 1990; O'Mahony & Ishii, 1987). A less elegant method is post hoc grouping of descriptors into putative categories constructed for the analysis (Halpern, 1987; Kelling and Halpern, 1988). When a post hoc analysis was done for the unrestricted verbal responses to four stimulus solutions, at least 80% of all descriptors for sodium saccharin, cherry Kool-Aid, and a urea, citric acid, and sodium saccharin mixture fell in a sweetness category. Membership in that category required that a spoken taste quality identification contain one or more of the words sweet, sugar, or saccharin. No other post hoc category contained as much as 25% of all responses, although a fifth of all descriptions of cherry Kool-Aid and of monosodium glutamate fit into a bitterness category; a similar proportion of monosodium glutamate responses also was appropriate for a food terms category that rarely accommodated responses to the other stimuli. These data certainly do not resolve the question of the importance or utility of using or avoiding limitations on taste quality description responses. Perhaps they suggest that the number of different descriptors employed by subjects need not specify how many taste perception categories are actually in use.

Both the number and nature of potential taste stimuli can also be specified by theoretical orientations. The orthodox rendering of the basic tastes theory indicates that there are but four types of taste stimuli, with NaCl, sucrose, HCl, and quinine (or a similar inorganic salt, simple sugar, acid, and alkaloid) being a complete set. This quartet, frequently called the "basic taste stimuli," is used with stunning regularity in investigations of taste psychophysics. In contrast, the objects in reference to which human taste perception presumably evolved, principally our food and beverages, contain a very much broader range of chemicals (see Section III,A,2,a). Direct tests demonstrate both that many of the food–derived chemicals not defined as basic taste stimuli are necessary to produce a mixture that has a taste similar to the foods we eat or drink, and that the primary taste stimuli, alone or in combinations, rarely taste like any known food or beverage (Section III,B,3,a).

E. Categories of Psychophysical Relationships

Psychophysics seeks to study, in valid, accurate, and precise ways, relationships between certain environmental energy patterns and behavioral responses to those patterns. Put slightly differently, "psychophysics is the science of the relationship between energy and the sensations it produces" (Burgard & Kuznicki, 1990). Not all environmental energies are of interest. Some will rather quickly kill or thoroughly incapacitate a particular organism, and therefore are not appropriate for experiments in which behavioral responses are the necessary outcomes. If the intensity of these lethal energies is sufficiently reduced, they may become nondamaging but also undetectable by the organism. Initial investigations, and some fatalities, may be unavoidable before lethal intensity ranges of energies are identified.

With seriously damaging events excluded, it still may be difficult to determine the appropriate energies for psychophysical studies of a sensory system. It is tempting to try all known, available, and safe energy patterns, with those that are effective at the lowest intensities designated as the natural and archetypal stimuli for hypothesized perceptual channels. However, this is not a cogent approach. These "best" stimuli may be absent from the usual environment of the organism at the necessary intensities, or may be present but invariant. Examination of the life-history of the organism and of the ecosystems in which it usually functions may provide valuable guides to appropriate energy patterns.

Some methods of psychophysical measurements have been in use for a long time whereas others are recent developments. All have been described in detail elsewhere, often with careful and deep analyses of their merits and problems. Neither instructions on how to carry out particular psychophysical methods nor dissections of the flaws and advantages of specific methods will be presented in this chapter. They are available in many other works (e.g., Atkinson, Herrnstein, Lindzey, & Luce, 1988; Boff, Kaufman, & Thomas, 1986; Burgard and Kuznicki, 1990; Meiselman and Rivlin, 1986; Piggott, 1988).

Psychophysical observations fall into a limited set of categories, and are made using relatively standard measurement procedures. There are three general categories: (1) thresholds, (2) judged intensity using energy levels different from (most typically but, in chemosensory domains, not necessarily, greater than) threshold levels, and (3) descriptions of the environmental energy patterns at nonthreshold levels. Limited aspects of taste intensity and taste description measurements were included in an earlier portion of this chapter that viewed taste as an ongoing process over time; gustatory descriptions were also a key factor in the prior discussions of theories on taste perception and of appropriate descriptive terms for taste. Category 1, thresholds, will be briefly discussed in this segment. Emphasis will be on temporal aspects of thresholds.

1. Thresholds

Thresholds conventionally represent a minimum detectable or describable intensity of a particular environmental energy pattern, or a minimum discriminable difference (Engen, 1986). *Detection* (sometimes called absolute) thresholds correspond to the minimum detectable condition, whereas *recognition* thresholds indicate the minimum intensity needed for consistent descriptions (Miller & Bartoshuk, 1991). *Dif-*

ferential thresholds exemplify the minimum discriminable difference between two successive intensity levels (Luce & Krumhansl, 1988). The required intensity levels for thresholds will differ depending on the species and condition of the organism, the sensory system under study, temporal factors such as the duration and rate of change of energy patterns, the particular wavelength, chemical, or spatial distribution of the stimuli, and the level and locus of measurement (e.g., receptor biophysics or biochemistry, neural activity, behavioral response).

a. Detection and Recognition Thresholds

Of what use are gustatory detection or recognition threshold data? Human psychophysical detection thresholds to solutions of single pure chemicals have proved valuable for diagnosing chemosensory disorders (Getchell et al., 1991). This suggests that such thresholds are connected in some fundamental way to the overall operation of chemosensory systems. In addition, psychophysical thresholds provide important sensitivity criteria for physiological and biophysical studies.

Threshold intensity levels measured using behavioral responses must be sufficient to activate not only receptor elements but also the entire sensory system. Nonetheless, it is not uncommon for psychophysical thresholds to be lower than those measured using neural or biophysical approaches. An evolutionary or adaptive perspective provides a reasonable explanation for the frequently greater sensitivity of psychophysical measures: a close fit should exist between the taste receptor apparatus of an organism and the neural processing that occurs in the gustatory regions of the nervous system of the organism. These intraorganismic input and processing connections, which were selected through evolutionary events, are likely to be much more effective than those between a sensory system and the measurements and analyses that an experimenter chooses to make.

Despite the uses of threshold measures just discussed, one must wonder if the intensity levels of detection or recognition thresholds are relevant to normal human perception? For chemosensory perception, the answer may be "yes" under some circumstances. This occurs because many of the chemical constituents of our food and drink are, until ingestion commences, either totally absent from our mouth or present at concentrations much below both recognition and detection thresholds. During drinking and eating, the intraoral concentrations rise. In doing so, they must pass through the threshold intensity regions. An important factor in relevance to perception is the rate at which the concentration change occurs. If the rate of movement through the threshold range is very swift, stimulation at threshold levels may be slight. On the other hand, an extremely slow increase in concentration within the threshold range could also result in ineffective chemosensory stimuli. Unfortunately, rates of intraoral concentration change seem to be unknown for normal eating or drinking.

Some information on concentration changes at the tongue surface has been obtained under laboratory conditions. Flowing liquid directly onto the human tongue produces quite rapid increases in concentration (Kelling & Halpern, 1986). For a solution of the synthetic sweetener sodium saccharin, the concentration of liquid flowing off of the tongue changes from that of purified water to 1.8 mM within 135 msec when a 2 mM sodium saccharin solution flows onto the tongue at velocity of 5 ml/sec. Because this velocity is within the range that occurs during normal human drinking, the measured rate of change in concentration on the tongue may approximate that which occurs during typical suction drinking (Halpern, 1991).

Interpretation of these psychophysical measurements requires another question to be answered: Are chemosensory environmental events with durations of less than 200 msec too brief for humans to detect or describe? Our studies with human observers found that 50-msec flows of 2 mM sodium saccharin, as well as other solutions, were both reliably detected and described qualitatively and quantitatively (Kelling & Halpern, 1988). Failures to detect that the solution was presented did occur more frequently than noted at longer durations, but the taste descriptions given, and the time intervals before they occurred, did not differ significantly from those at longer stimulus durations. However, judged intensity did increase significantly with stimulus duration throughout the range of 50- to 2000-msec durations.

As already suggested, at the beginning of a sip, although perhaps only for the first sip when a series of sips occurs (Halpern, 1986b), concentrations in the mouth may increase to, and pass through, the taste detection and recognition threshold range. Although the dwell time within the threshold concentration zone may be brief, humans respond so well to short-duration taste stimuli that detection of some change in taste, and then characterization of what the liquid tasted like and how intense it was, may occur. Taste reaction time measurements indicate that 0.5 sec or more could separate the concentration change in the mouth from the perceived characterization of taste identity and intensity (Kelling & Halpern, 1988). Perhaps this hypothesized series of intraoral concentration changes and associated sensory and perceptual events is real. If so, then "taste flashes" with durations of about 50 msec, which were once described as a laboratory concoction that was unrelated to normal human experience (Kelling & Halpern, 1983), may be a useful initial component of human tasting.

The initial sip of a sequence was stipulated in the preceding analysis because intraoral concentrations do not immediately decrease once a sip and swallow (or spit) are completed. Instead, a substantial residue exists in the mouth for durations that greatly exceed the typical time intervals between sips (Halpern, 1986a) It follows that concentrations in a detection or recognition threshold region are likely to be ephemeral during drinking, with higher concentrations present at most times.

Eating of more-or-less solid foods may present a different situation. Chewing and manipulation of foods will allow relatively gradual transfer of some constituents into saliva and thence to gustatory receptor surfaces, although the rate would vary with the nature of the food. Time within detection or threshold ranges could be longer than during the drinking of liquids, thus extending the duration of threshold-level sensory input. Thermal and some mechanical aspects of foods, which can affect not only intraoral somatosensory systems but also taste receptors, might be perceived long before chemosensory responses (Yamamoto & Kawamura, 1981, 1984). Unfortunately, the actual time course of both intraoral chemosensory events and taste perception during the ingestion and mastication of food is largely unknown.

b. Detection versus Recognition Thresholds

In general, detection thresholds are defined by an intensity level that permits a reliable distinction between the presence, or the total absence, of a particular environmental energy. The measurement operations are two-category procedures that neither require nor request any perceptual description. At detection threshold intensity, no consistent description of the energy, and no characterization of the degree of intensity, may be possible. More intense stimuli are often required to permit descriptions. The latter intensity defines the recognition threshold.

The distinction between detection and recognition thresholds has long been a part of the psychophysical tradition. An implicit assumption is that when sufficient energy is available, not only will consistent descriptions become possible but also these descriptions will be the "correct" descriptions. This characterization may be valid for hearing and seeing. However, it does not fit taste well. A number of instances exist in which a reliable taste description of a chemical at a particular concentration changes to a different, but also consistent description, at a higher concentration (see Pfaffmann, 1959). For example, aqueous solutions of NaCl between about 10 and 30 mM are described as sweet, but at concentrations of 50 mM and above receive a description of salty. LiCl solutions are also largely sweet at concentrations below 10 mM, predominantly sour between 10 and 30 mM, and evoke equal reports of salty and sour at 50 mM (Dzendolet & Meiselman, 1967). The taste of KCl solutions alters even more dramatically, with reports of sweet or bitter prevalent below 10 mM, bitterness predominant around 20 mM, and salty or sometimes sour around 50 mM (Pfaffmann et al., 1971). Such changes may occur several times as increasingly more concentrated solutions of a single molecule are presented. Which concentration identifies the recognition threshold range? A common practice is to take the description given for the relatively high concentrations that are often used in research as the canonical ones. However, during normal tasting, concentrations may typically rise from subthreshold levels to those well above the "official" recognition threshold at a rate that permits responses at the intervening concentrations. Then, either there is a series of valid taste recognition thresholds, or the lowest concentration that elicits a reliable description should be identified as the taste recognition threshold.

IV. OVERVIEW

The design, execution, analysis, and interpretation of psychophysical investigations of taste are strongly influenced and constrained by an array of assumptions and theoretical positions. One pervasive model, the basic tastes theory, has especially strong influences. Its set of four primary stimulus chemicals, and four independent and immiscible taste receptor types and tightly linked perceptual categories, are both acknowledged and largely accepted by many investigators. A consequence is that studies are sometimes designed to discover the number of taste primaries, but less often attempt to test the validity of the concept of taste primaries.

Theoretical positions can also denote both the diversity and the attributes of potential taste stimuli. The traditional interpretation of the basic tastes theory specifies that four species of stimuli, NaCl, sucrose, HCl, and quinine, or a similar inorganic salt, simple sugar, acid, and alkaloid, are a full compilation. The components of this quadrate gustatory universe, commonly described as the basic taste stimuli, are customarily employed, often to the exclusion of all other chemicals, in investigations of taste psychophysics. However, the items in association to which human taste perception is surmised to have evolved, our food and beverages, present a considerably more expansive range of chemicals. Specific studies have shown both that many food-derived chemicals that are excluded from the basic taste stimuli list are indispensable for mixtures that taste similar to what we eat or drink, and that the primary taste stimuli, singly or combined, are seldom perceived to taste like any naturally consumed substances.

At the level of taste perception, basic tastes theory-derived studies usually permit subjects to report only four descriptions: bitter, salty, sour, and sweet. Not infrequently, subjects are taught to use these words in the desired manner, that is, bitter for quinine, salty for NaCl, sour for HCl, and sweet for sucrose, if their initial descriptions are incorrect or confused. I must note that adult subjects will, in the absence of constraints or correction, freely use some of these basic taste terms for aqueous solutions of single molecules, for example, sweet for sucrose and salty for NaCl. However, when permitted, subjects will also use adjective compounds containing not only basic taste terms such as sweet–salty but also compounds that include food object words, for example, sweet–apple. Aqueous solutions of some stimuli, for example monosodium glutamate, often evoke descriptors such as soapy.

Eating, drinking, and consequently tasting are dynamic processes extending over time. In contrast, psychophysical measures are traditionally taken only once during the presentation of a stimulus. When time-dependent measures are used for taste perception, a fairly rich structure is revealed. For examination of those changes in perceived taste intensity or taste quality descriptions that correspond to the temporal constraints of normal ingestion and taste-dependent decisions, stimulus durations of less than 5 sec, and perceptual measurements about 10 times/sec, are appropriate.

Thresholds are a common but problematic measure in studies of taste perception. The relevance of taste thresholds to normal taste-dependent behavior has often been questioned. At the beginning of ingestion, threshold regions will necessarily be traversed. Because 50-msec stimulus durations are sufficient to permit consistent judgments of taste intensity and reliable description of taste quality, it may be that the time intervals during which taste stimuli are in the threshold range are sufficient to provide useful gustatory information.

Overall, there remains much to be learned and understood concerning the psychophysics of taste. Although current theoretical models have some utility, neither the basic tastes approach nor pattern theories of taste map very well into human eating and drinking behaviors or taste-dependent decisions. Hopes of revealing universal laws of tasting, reliance on physical rather than biological definitions of stimulus simplicity, and excessively facile movement between quite different levels of analysis have often led to parochial, contentious, and convoluted gustatory hypotheses, and nonilluminating experiments. One possible solution is an open and nondogmatic approach to the complexity of the psychophysics of taste and of taste perception, with a constant albeit sometimes distant validity check provided by behavioral responses to the beverages and foods that humans choose and consume or reject.

Acknowledgments

I thank Jeannine F. Delwiche, Kathleen M. Dorries, Robert P. Erickson, Herbert L. Meiselman, Thomas Neuhaus, Michael O'Mahony, and Susan S. Schiffman for critical comments on earlier drafts of this chapter.

References

American heritage electronic dictionary (1992). (Version 1.1, college/professional ed.). Novato, CA: WordStar International.

American heritage electronic dictionary (1993). [CD-ROM] (Deluxe 3rd ed.). Novato, CA: StarPress.

Amerine, M. A., Pangborn, R. M., & Roessler, E. B. (1965). Principles of sensory evaluation of food. New York: Academic Press.

Andersen, H. T. (1970). Problems of taste specificity. In G. E. W. Wolstenholme & J. Knight (Eds.), Taste and smell in vertebrates (pp. 71–82). London: Churchill.

Atkinson, R. C., Herrnstein, R. J., Lindzey, G., & Luce, R. D. (Eds.). (1988). Steven's handbook of experimental psychology: Vol. 1. Perception and motivation (2nd ed.), New York: Wiley.

Ayya, N., & Lawless, H. T. (1992). Quantitative and qualitative evaluation of high-intensity sweeteners and sweetener mixtures. Chemical Senses, 17, 245–259.

Backus, J. (1977). The acoustical foundations of music (2nd ed.). New York: Norton.

Barlow, H. B., & Mollon, J. D. (Eds.) (1982). The senses. Cambridge: Cambridge Univ. Press.

Bartoshuk, L. M. (1980). Sensory analysis of the taste of NaCl. In M. R. Kare, M. J. Fregley, & R. A. Bernard (Eds.), Biological and behavioral aspects of salt intake (pp. 83–98). New York: Academic Press.

Bartoshuk, L. M. (1987). Is sweetness unitary? An evaluation of the evidence for multiple sweets. In J. Dobbing (Ed.), Sweetness (pp. 33–47). London: Springer-Verlag.

Bartoshuk, L. M. (1988). Taste. In R. C. Atkinson, R. J. Herrnstein, G. Lindzey, & R. D. Luce (Eds.), Steven's handbook of experimental psychology Vol. 1. Perception and motivation (2nd ed.) (pp. 461–499). New York: Wiley.

Bartoshuk, L. M., & Gent, J. F. (1985). Taste mixtures: An analysis of synthesis. In D. W. Pfaff (Ed.), Taste, olfaction, and the central nervous system. New York: Rockefeller Univ. Press.

Bartoshuk, L. M., & Marks, L. E. (1986). Ratio scaling. In H. L. Meiselman & R. S. Rivlin (Eds.), Clinical measurement of taste and smell (pp. 50–65). New York: Macmillan.

Beauchamp, G. K., Cowart, B. J., & Schmidt, H. J. (1991). Development of chemosensory sensitivity and preference. In T. V. Getchell, R. L. Doty, L. M. Bartoshuk, & J. B. Snow, Jr. (Eds.), *Smell and taste in health and disease* (pp. 405–416). New York: Raven.

Beets, M. G. J. (1978). *Structure-activity relationships in human chemoreception.* London: Applied Science.

Boff, K. R., Kaufman, L., & Thomas, J. P. (Eds.). (1986). *Handbook of perception and human performance: Vol. I. Sensory processes and perception.* New York: Wiley.

Bolanowski, S. J., Jr., & Gescheider, A. G. (Eds.). (1991). *Ratio scaling of psychological magnitude.* Hillsdale, NJ: Erlbaum.

Bonnans, S., & Noble, A. C. (1993). Effect of sweetener type and acid level on temporal perception of sweetness, sourness and fruitiness. *Chemical Senses, 18,* 273–283.

Boring, E. G. (1942). *Sensation and perception in the history of experimental psychology.* New York: Appleton-Century-Crofts.

Boring, E. G. (1950). *A history of experimental psychology* (2nd ed.). New York: Appleton-Century-Crofts.

Boudreau, J. C. (Ed.). (1979). *Food taste chemistry* (ACS symposium 115). Washington, DC: American Chemical Society.

Boughter, J. D., Jr., & Whitney, G. (1993). Human taste threshold for sucrose octaacetate. *Chemical Senses, 18,* 445–448.

Brand, J. G., Teeter, J. H., Kumazawa, T., Taufiqul, H., & Bayley, D. L. (1991). Transduction mechanisms for the taste of amino acids. *Physiology & Behavior, 49,* 863–868.

Breslin, P. A. S. (1996). Interactions among salty, sour and bitter compounds. *Trends in Food Science and Technology, 7,* 390–399.

Breslin, P. A. S., & Beauchamp, G. K. (1994). Strong acids are indiscriminable at equal pH. *Chemical Senses, 19,* 447.

Breslin, P. A. S., Gilmore, M. M., Beauchamp, G. K., & Green, B. G. (1993). Psychophysical evidence that oral astringency is a tactile sensation. *Chemical Senses, 18,* 405–417.

Breslin, P. A. S., Beauchamp, G. K., & Pugh, E. N., Jr. (1996). Monogeusia for fructose, glucose, sucrose, and maltose. *Perception and Psychophysics, 58,* 327–341.

Breslin, P. A. S., Kemp, S., and Beauchamp, G. K. (1994). Single sweetness signal. *Nature, 369,* 447–448.

Burgard, D. R., & Kuznicki, J. T. (1990). *Chemometrics.* Boca Raton, FL: CRC Press.

Cagan, R. H., Brand, J. G., Morris, J. A., & Morris, R. W. (1978). Monellin, a sweet-tasting protein. In J. H. Shaw & G. G. Roussos (Eds.), Proceeding 'sweeteners and dental caries'. *Special supplement to feeding, weight & obesity abstracts.* pp. 327–338.

Capranica, R. R. (1992). The untuning of the tuning curve: Is it time? *The Neurosciences, 4,* 401–408.

Crocker, E. C. (1945). *Flavor.* New York: McGraw-Hill.

Delconte, J. D., Kelling, S. T., & Halpern, B. P. (1992). Speed and consistency of human decisions to swallow or spit sweet and sour solutions. *Experientia, 48,* 1106–1109.

Delcour, J. A., Vandenberghe, M. M., Corten, P. F., & Dondeyne, P. (1984). Flavor thresholds of polyphenolics in water. *American Journal of Enology and Viticulture, 35,* 134–136.

DeSimone, J. A. (1991). Transduction in taste receptors. *Nutrition, 7,* 146–149.

Dowling, J. E. (1987). *The retina: An approachable part of the brain.* Cambridge, MA: The Belknap Press of Harvard University.

Dubner, R., Sessle, B. J., & Storey, A. T. (1978). *The neural basis of oral and facial function.* New York: Plenum.

Dzendolet, E., & Meiselman, H. L. (1967). Gustatory quality changes as a function of solution concentration. *Perception & Psychophysics, 2,* 29–33.

Engen, T. (1986). Classical psychophysics: Humans as sensors. In H. L. Meiselman & R. S. Rivlin (Eds.), *Clinical measurement of taste and smell* (pp. 39–49). New York: Macmillan.

Erickson, R. P. (1963). Sensory neural patterns and gustation. In Y. Zotterman (Ed.), *Olfaction and taste* (pp. 205–213). London: Pergamon.

Erickson, R. P. (1977). The role of 'primaries' in taste research. In J. Le Magnen & P. Mac Leod (Eds.), *Olfaction and taste VI* (pp. 369–376). London: Information Retrieval.

Erickson, R. P. (1984). On the neural bases of behavior. *American Scientist, 72,* 233–240.

Erickson, R. P. (1985). Definitions: A matter of taste. In D. W. Pfaff (Ed.), *Taste, olfaction, and the central nervous system* (pp. 129–150). New York: Rockefeller Univ. Press.

Erickson, R. P., & Covey, E. (1980). On the singularity of taste sensations: What is a primary taste? *Physiology & Behavior, 25,* 527–533.

Erickson, R. P., Priolo, C. V., Warwick, Z. S., & Schiffman, S. S. (1990). Synthesis of tastes other than the 'primaries': Implications for neural coding theories and the concept of 'suppression'. *Chemical Senses, 15,* 495–504.

Erickson, R. P., Rodgers, J. L., & Sarle, W. S. (1993). Statistical analysis of neural organization. *Journal of Neurophysiology, 70,* 2289–2300.

Faurion, A. (1987a). MSG as one of the sensitivities within a continuous taste space: Electrophysiological and psychophysical studies. In Y. Kawamura & M. R. Kare (Eds.), *Umami: A basic taste* (pp. 387–408). New York: Dekker.

Faurion, A. (1987b). Physiology of the sweet taste. *Progress in Sensory Physiology, 8,* 129–201.

Faurion, A. (1993). The physiology of sweet taste and molecular receptors. In M. Mathlouthi, J. A. Kanter, & G. G. Birch (Eds.), *Sweet-taste chemoreception* (291–315). London: Elsevier.

Faurion, A. (1994). Structure and dimensions of taste sensory space, central and peripheral data. In K. Kurihara, N. Suzuki, & H. Ogawa (Eds.), *Olfaction and taste XI* (pp. 301–304). Tokyo: Springer-Verlag.

Faurion, A., Saito, S., & Mac Leod, P. (1980). Sweet taste involves several distinct receptor mechanisms. *Chemical Senses, 5,* 107–121.

Fennema, O. R. (Ed.). (1985). *Food chemistry* (2nd ed.). New York: Dekker.

Finger, T. E., & Silver, W. L. (Eds.). (1987). *Neurobiology of taste and smell.* New York: Wiley.

Frijters, J. E. R., Blauw, Y. H., & Vermatt, S. H. (1982). Incidental training in the triangular method. *Chemical Senses, 7,* 63–69.

Fuke, S., & Konosu, S. (1991). Taste-active components in some foods: A review of Japanese research. *Physiology & Behavior, 49,* 863–868.

Ganzevles, P. G. J., & Kroeze, J. H. A. (1987). The sour taste of acids. The hydrogen ion and the undissociated acid as sour agents. *Chemical Senses, 12,* 563–567.

Getchell, T. V., Doty, R. L., Bartoshuk, L. M., & Snow, J. B., Jr. (Eds.). (1991). *Smell and taste in health and disease.* New York: Raven.

Gibson, J. J. (1966). *The senses considered as perceptual systems.* Boston: Houghton Mifflin.

Gillette, M. (1985). Flavor effects of sodium chloride. *Food Technology, 39,* 47–52 and 56.

Gouras, P., & Zrenner, E. (1981). Color vision: A review from a neurophysiological perspective. *Progress in Sensory Physiology, 1,* 139–179.

Green, B. G., & Lawless, H. T. (1991). The psychophysics of somatosensory chemoreception in the nose and mouth. In T. V. Getchell, R. L. Doty, L. M. Bartoshuk, & J. B. Snow, Jr. (Eds.), *Smell and taste in health and disease* (pp. 235–253). New York: Raven.

Gulick, W. L., Gescheider, G. A., & Frisina, R. D. (1989). *Hearing: physiological acoustics, neural coding, and psychoacoustics.* New York: Oxford Univ. Press.

Halpern, B. P. (1973). The other senses. In B. B. Wolman (Ed.), *Handbook of general psychology* (pp. 382–384). Englewood Cliffs, NJ: Prentice-Hall.

Halpern, B. P. (1985). Time as a factor in gustation: Temporal patterns of taste stimulation and response. In D. W. Pfaff (Ed.), *Taste, olfaction, and the central nervous system* (pp. 181–209). New York: Rockefeller Univ. Press.

Halpern, B. P. (1986a). Constraints imposed on taste physiology by human taste reaction time data. *Neuroscience and Biobehavioral Reviews, 10,* 135–151.

Halpern, B. P. (1986b). What to control in studies of taste. In H. L. Meiselman & R. S. Rivlin (Eds.), *Clinical measurement of taste and smell* (pp. 126–153). New York: Macmillan.

Halpern, B. P. (1987). Human judgments of msg tastes: Quality and reaction times. In Y. Kawamura & M. R. Kare (Eds.), *Umami: A basic taste* (327–354). New York: Dekker.

Halpern, B. P. (1991). More than meets the tongue: Temporal characteristics of taste intensity and quality. In H. T. Lawless & B. P. Klein (Eds.), *Sensory science theory and applications in foods* (pp. 37 – 105). New York: Dekker.

Halpern, B. P. (1994). Temporal patterns of perceived tastes differ from liquid flow at the tongue. In K. Kurihara, N. Suzuki, & H. Ogawa (Eds.), *Olfaction and taste XI* (pp. 297 – 300). Tokyo: Springer-Verlag.

Halpern, B. P., Kelling, S. T., Davis, J., Dorries, K. M., Haq, A., & Meltzer, J. S. (1992). Effects of amiloride on human taste responses to NaCl: Time-intensity and taste quality descriptor measures. [Abstract]. *Chemical Senses, 17,* 637.

Halpern, B. P., Meltzer, J. S., & Darlington, R. D. (1993). Effects of amiloride on tracked taste intensity and on taste quality descriptions of NaCl: Individual differences and dose-response effects. [Abstract]. *Chemical Senses, 18,* 566.

Harnad, S. (1987). Psychophysical and cognitive aspects of categorical perception: A critical overview. In S. Harnad (Ed.), *Categorical perception: the groundwork of cognition* (pp. 1 – 25). New York: Cambridge Univ. Press.

Hettinger, T. P., & Frank, M. E. (1992). Information processing in mammalian gustatory systems. *Current Opinion in Neurobiology, 2,* 469 – 478.

Hettinger, T. P., Myers, W. E., & Frank, M. E. (1990). Role of olfaction in perception of non-traditional 'taste' stimuli. *Chemical Senses, 15,* 755 – 760.

Hoffmann, R., & Torrence, V. (1993). *Chemistry imagined: Reflections on science.* Washington, DC: Smithsonian.

Holway, A. H., and Hurvich, L. M. (1937). Differential gustatory sensitivity to salt. *American Journal of Psychology, 49,* 37 – 48.

Inglett, G. E. (1978). Potential intense sweeteners of natural origin. In J. H. Shaw & G. G. Roussos (Eds.), *Proceeding "sweeteners and dental caries" Special Supplement to Feeding, Weight & Obesity Abstracts* (pp. 311 – 325).

Ishii, R., & O'Mahony, M. (1987). Defining a taste by a single standard: Aspects of salty and umami taste. *Journal of Food Science, 52,* 1405 – 1409.

Ishii, R., & O'Mahony, M. (1991). Use of multiple standards to define gustatory characteristics for descriptive analysis. *Journal of Food Science, 56,* 838 – 842.

Kander, S. (1954). *The new settlement cook book* (revised and enlarged, p. 527). New York: Simon and Schuster.

Kato, H., Rhue, M. R., & Nishimura, T. (1989). Role of free amino acids and peptides in food taste. In R. Teranishi, R. G. Butler, and F. Shahidi (Eds.), *Flavor chemistry: Trends and developments* (ACS Symposium 388, pp. 158 – 174). Washington, DC: American Chemical Society.

Kelling, S. T., & Halpern, B. P. (1983). Taste flashes: Reaction times, intensity, and quality. *Science, 219,* 412 – 414.

Kelling, S. T., & Halpern, B. P. (1986). Physical characteristics of open flow and closed flow taste delivery apparatus. *Chemical Senses, 11,* 89 – 104.

Kelling, S. T., & Halpern, B. P. (1987). Taste judgments and gustatory stimulus duration: Simple taste reaction times. *Chemical Senses, 12,* 543 – 562.

Kelling, S. T., & Halpern, B. P. (1988). Taste judgments and gustatory stimulus duration: Taste quality, taste intensity, and reaction time. *Chemical Senses, 13,* 559 – 586.

Kinnamon, S. C., & Getchell, T. V. (1991). Sensory transduction in olfactory receptors neurons and gustatory receptor cells. In T. V. Getchell, R. L. Doty, L. M. Bartoshuk, & J. B. Snow, Jr. (Eds.), *Smell and taste in health and disease* (pp. 145 – 172). New York: Raven.

Kling, J. W., & Riggs, L. A. (Eds.). (1971). *Woodworth & Schlosberg's experimental psychology* (3rd ed.). New York: Holt, Rinehart, and Winston.

Konosu, S. (1979). The taste of fish and shellfish. In J. C. Boudreau (Ed.), *Food taste chemistry* (ACS Symposium 115, pp. 185 – 203). Washington, DC: American Chemical Society.

Konosu, S., Yamaguchi, K., & Hayashi, T. (1987). Role of extractive components of boiled crab in producing the characteristic flavor. In Y. Kawamura & M. R. Kare (Eds.), *Umami: A basic taste* (pp. 235 – 253). New York: Dekker.

Köster, E. P. (1981). Time and frequency analysis: A new approach to the measurement of some less-well-known aspects of food preferences? In J. Solms & R. L. Hall (Eds.), *Criteria of food acceptance* (pp. 240–252). Zurich, Switzerland, Forster.

Kroeze, J. H. A. (1990). The perception of complex taste stimuli. In R. L. McBride & H. J. H. MacFie (Eds.), *Psychological basis of sensory evaluation* (pp. 41–68). London: Elsevier.

Kroeze, J. H. A., & Bartoshuk, L. M. (1985). Bitterness suppression as revealed by split-tongue taste stimulation in humans. *Physiology & Behavior, 35,* 779–783.

Kurihara, K. (1987). Recent progress in the taste receptor mechanism. In Y. Kawamura & M. R. Kare (Eds.), *Umami: A basic taste* (pp. 3–39). New York: Dekker.

Kurihara, Y. (1992). Characteristics of antisweet substances, sweet proteins, and sweetness-inducing proteins. *Critical Reviews in Food Science and Nutrition, 32,* 231–252.

Kurihara, Y. (1994). Structures and functions of antisweetness substances, sweetness-inducing substances, and sweet proteins. In K. Kurihara, N. Suzuki, & H. Ogawa (Eds.), *Olfaction and taste XI* (pp. 242–245). Tokyo: Springer-Verlag.

Kuznicki, J. T., & Cardello, A. V. (1986). Psychophysics of single taste papillae. In H. L. Meiselman & R. S. Rivlin (Eds.), *Clinical measurement of taste and smell* (pp. 200–225). New York: Macmillan.

Lawless, H. T. (1979). Evidence for neural inhibition in bittersweet taste mixtures. *Journal of Comparative and Physiological Psychology, 93,* 538–547.

Lawless, H. T. (1991). Gustatory psychophysics. In T. E. Finger & W. L. Silver (Eds.), *Neurobiology of taste and smell* (pp. 401–420). New York: Wiley.

Lawless, H. T., Corrigan, C. J., and Lee, C. B. (1994). *Interaction of astringent substances. Chemical Senses, 19,* 141–154.

Lawless, H. T., & Clark, C. C. (1992). Psychological biases in time-intensity scaling. *Food Technology, 46,* 81, 84–86, and 90.

Lawless, H. T., & Klein, B. P. (1991). (Eds.). *Sensory Science Theory and Applications in Foods.* New York: Dekker.

Levine, M. W., & Shefner, J. M. (1991). *Fundamentals of sensation and perception* (2nd ed.). Belmont, CA: Brooks/Cole.

Lindsay, R. C. (1985). Flavors. In O. R. Fennema (Ed.), *Food chemistry* (2nd ed., pp. 585–626). New York: Dekker.

Luce, R. D., & Krumhansl, C. L. (1988). Measurement, scaling, and psychophysics. In R. C. Atkinson, R. J. Herrnstein, G. Lindzey, & R. D. Luce (Eds.), *Steven's handbook of experimental psychology: Vol. 1. Perception and Motivation* (2nd ed., pp. 3–74). New York: Wiley.

McBurney, D. M. (1974). Are there primary tastes for man? *Chemical Senses and Flavor, 1,* 17–28.

McBurney, D. M. (1984). Taste and olfaction: Sensory discrimination. In J. M. Brookhard & V. B. Mountcastle (Eds.), *Handbook of physiology: The nervous system. III. Sensory processes* (pp. 1067–1086). Baltimore: Williams & Wilkins.

McBurney, D. M. (1986). Taste, smell, and flavor terminology: Taking the confusion out of fusion. In H. L. Meiselman & R. S. Rivlin (Eds.), *Clinical measurement of taste and smell* (pp. 117–125). New York: Macmillan.

Medin, D. L. (1989). Concepts and conceptual structure. *American Psychologist, 44,* 1469–1481.

Meilgaard, M. C., Civille, G. V., & Carr, B. T. (1991). *Sensory evaluation techniques* (2nd ed.). Boca Raton, FL: CRC Press.

Meiselman, H. L. (1988). Consumer studies of food habits. In J. R. Piggott (Ed.), *Sensory analysis of foods* (2nd ed., pp. 267–334). London: Elsevier.

Meiselman, H. L. (1992a). Methodology and theory in human eating research. *Appetite, 19,* 49–55.

Meiselman, H. L. (1992b). Obstacles to studying real people eating real meals in real situations. *Appetite, 19,* 84–86.

Meiselman, H. L. (1993). Critical evaluation of sensory techniques. *Food Quality and Preference, 4,* 33–40.

Meiselman, H. L., & Rivlin, R. S. (Eds.). (1986). *Clinical measurement of taste and smell.* New York: Macmillan.

Miller, I. J., & Bartoshuk, L. M. (1991). Taste perception, taste bud distribution, and spatial relationships. In T. V. Getchell, R. L. Doty, L. M. Bartoshuk, & J. B. Snow, Jr. (Eds.), *Smell and taste in health and disease* (pp. 205–233). New York: Raven.

Mollon, J. D. (1982). Colour vision and colour blindness. In H. B. Barlow & J. D. Mollon (Eds.) *The senses* (pp. 165–191). Cambridge: Cambridge Univ. Press.

Monti-Bloch, L., & Grosser, B. I. (1991). Effect of putative pheromones on the electrical activity of the human vomeronasal organ and olfactory epithelium. *Journal of Steroid Biochemistry and Molecular Biology, 39,* 573–582.

Moore, B. C. J. (1982). *An introduction to the psychology of hearing.* London: Academic Press.

Moskowitz, H. R. (1984). Sensory analysis, product modeling, and product optimization. In G. Charalambous (Ed.), *Analysis of foods and beverages: Modern techniques* (pp. 13–67). London: Academic Press.

Nishimura, T., & Kato, H. (1988). Taste of free amino acids and peptides. *Food Reviews International, 4,* 175–194.

Noble, A. C. (1990). Bitterness and astringency in wine. In R. L. Rouseff (Ed.), *Bitterness in foods and beverages* (pp. 145–158). Amsterdam: Elsevier.

Oakley, B. (1986). Basic taste physiology: Human perspectives. In H. L. Meiselman & R. S. Rivlin (Eds.), *Clinical measurement of taste and smell* (pp. 5–18). New York: Macmillan.

O'Mahony, M. (1986). *Sensory evaluation of food.* New York: Dekker.

O'Mahony, M. (1990). Cognitive aspects of difference testing and descriptive analysis: Criterion variation and concept formation. In R. L. McBride & H. J. H. MacFie (Eds.), *Psychological basis of sensory evaluation* (pp. 117–139). London: Elsevier.

O'Mahony, M. (1991). Descriptive analysis and concept alignment. In H. T. Lawless & B. P. Klein (Eds.), *Sensory science theory and applications in foods* (pp. 223–267). New York: Dekker.

O'Mahony, M., Atassi-Sheldon, S., Rothman, L., & Murphy-Ellison, T. (1983). Relative singularity/mixedness judgements for selected taste stimuli. *Physiology & Behavior, 31,* 749–755.

O'Mahony, M., & Buteau, L. (1982). Taste mixtures: Can the components be readily identified? *IRCS Medical Sciences: Alimentary System; Biochemistry; Biomedical Technology; Dentistry and Oral Biology; The Eye; Physiology; Psychology and Psychiatry; Social and Occupational Medicine, 10,* 109–110.

O'Mahony, M., Goldenberg, M., Stedmon, J., & Alford, J. (1979). Confusion in the use of the adjectives 'sour' and 'bitter'. *Chemical Senses and Flavour, 4,* 301–318.

O'Mahony, M., & Ishii, R. (1987). The umami taste concept: Implications for the dogma of four basic tastes. In Y. Kawamura & M. R. Kare (Eds.), *Umami: A basic taste* (pp. 75–93). New York: Dekker.

O'Mahony, M., Rothman, L., Ellison, T., Shaw, D., & Buteau, L. (1990). Taste descriptive analysis: Concept formation, alignment and appropriateness. *Journal of Sensory Studies, 5,* 71–103.

Pangborn, R. M. (1981). A critical review of threshold, intensity and descriptive analyses in flavor research. In *Flavour '81* (pp. 3–32). Berlin: de Gruyter.

Pfaffmann, C. (1959). The sense of taste. In J. Field (Ed.), *Handbook of physiology. Section 1: Neurophysiology, Vol. I* (pp. 507–533). Washington, DC: American Physiological Society.

Pfaffmann, C., Bartoshuk, L. M., & McBurney, D. H. (1971). Taste psychophysics. In L. M. Beidler (Ed.), *Handbook of sensory physiology: Vol IV, Chemical senses. Part 2. Taste* (pp. 75–101). Berlin: Springer-Verlag.

Pickles, J. O. (1988). *An introduction to the physiology of hearing* (2nd ed.). London: Academic Press.

Piggott, J. R. (Ed.). (1988). *Sensory analysis of foods* (2nd ed.). London: Elsevier.

Plattig, K.-H. (1991). Gustatory evoked brain potentials in humans. In T. V. Getchell, R. L. Doty, L. M. Bartoshuk, & J. B. Snow, Jr. (Eds.), *Smell and taste in health and disease* (pp. 277–286). New York: Raven.

Powers, J. J. (1988). Current practices and applications of descriptive methods. In J. R. Piggott (Ed.), *Sensory analysis of foods* (2nd ed., pp. 187–265). London: Elsevier.

Prescott, J., Allen, S., & Stephens, L. (1993). Interactions between chemical irritation, taste and temperature. *Chemical Senses, 18,* 389–404.

Pugh, E. N. (1988). Vision: Physics and retinal physiology. In R. C. Atkinson, R. J. Herrnstein,

G. Lindzey, & R. D. Luce (Eds.), *Steven's handbook of experimental psychology: Vol. 1, Perception and motivation* (2nd ed., pp. 75–163). New York: Wiley.

Redlinger, P. A., & Setser, C. S. (1987). Sensory quality of selected sweeteners: Aqueous and lipid model systems. *Journal of Food Science, 52,* 451–454.

Robichaud, J. L., & Noble, A. C. (1990). Astringency and bitterness of selected phenolics in wine. *Journal of the Science of Food and Agriculture, 53,* 343–353.

Robins, R. J., Rhodes, M. J. C., Parr, A. J., & Walton, N. J. (1990). The biosynthesis of bitter compounds. In R. L. Rouseff (Ed.), *Bitterness in food and beverages* (pp. 49–79). Amsterdam: Elsevier.

Rosch, E., Mervis, C. B., Gray, W. D., Johnson, D., & Boyes-Braem, P. (1976). Basic objects in natural categories. *Cognitive Psychology, 8,* 382–439.

Rouseff, R. L. (1990). Bitterness in food products: An overview. In R. L. Rouseff (Ed.), *Bitterness in foods and beverages* (pp. 1–14). Amsterdam: Elsevier.

Rubico, S. M., & McDaniel, M. R. (1992). Sensory evaluation of acids by free-choice profiling. *Chemical Senses, 17,* 273–289.

Schiffman, S. S., & Erickson, R. P. (1980). The issue of primary tastes versus a taste continuum. *Neuroscience & Biobehavioral Reviews, 4,* 109–117.

Schiffman, S. S., & Erickson, R. P. (1993). Psychophysical insights into transduction mechanisms and neural coding. In S. A. Simon & S. D. Roper (Eds.), *Mechanisms of taste transduction* (pp. 395–424). Boca Raton, FL: CRC Press.

Schiffman, S. S., & Gill, J. M. (1987). Psychophysical and neurophysiological taste responses to glutamate and purinergic compounds. In Y. Kawamura & M. R. Kare (Eds.), *Umami: A basic taste* (pp. 271–288). New York: Dekker.

Schiffman, S. S., Hopfinger, A. J., & Mazur, R. H. (1986). The search for receptors that mediate sweetness. In M. Conn (Ed.), *The receptors, Vol. IV* (pp. 315–377). New York: Academic Press.

Schiffman, S. S., & Sattely-Miller, E. A. (1994). Perception of monosodium glutamate in water and in foods by young and elderly subjects. In K. Kurihara, N. Suzuki, & H. Ogawa (Eds.), *Olfaction and taste XI* (pp. 348–352). Tokyo: Springer-Verlag.

Scoble, M. J. (1992). *The lepidoptera: Form, function, and diversity.* New York: Oxford Univ. Press.

Scott, T. R., & Plata-Salaman, C. R. (1991). Coding of taste quality. In T. V. Getchell, R. L. Doty, L. M. Bartoshuk, & J. B. Snow, Jr. (Eds.), *Smell and taste in health and disease* (pp. 345–368). New York: Raven.

Sekuler, R., & Blake, R. (1985). *Perception.* New York: Knopf.

Sekuler, R., & Blake, R. (1994). *Perception* (3d ed.). New York: McGraw-Hill.

Silver, W. L., & Finger, T. E. (1991). The trigeminal system. In T. V. Getchell, R. L. Doty, L. M. Bartoshuk, & J. B. Snow, Jr. (Eds.), *Smell and taste in health and disease* (pp. 97–108). New York: Raven.

Simon, H. (1993). Cognitive Studies Program Colloquium at Cornell University, April 23, 1993, Ithaca, New York.

Simon, S. A., & Roper, S. D. (Eds.) (1993). *Mechanisms of taste transduction.* Boca Raton, FL: CRC Press.

Simon, S. A., & Wang, Y. (1993). Chemical responses of lingual nerves and lingual epithelia. In S. A. Simon & S. D. Roper (Eds.), *Mechanisms of taste transduction* (pp. 225–272). Boca Raton, FL: CRC Press.

Smith, D. V. (1985). Brainstem processing of gustatory information. In D. W. Pfaff (Ed.), *Taste, olfaction, and the central nervous system* (pp. 151–177). New York: Rockefeller Univ. Press.

Smith, D. V., & Frank, M. E. (1993). Sensory coding by peripheral taste fibers. In S. A. Simon & S. D. Roper (Eds.), *Mechanisms of taste transduction* (pp. 295–337). Boca Raton, FL: CRC Press.

Snow, J. B., Doty, R. L., Bartoshuk, L. M., & Getchell, T. V. (1991). Categorization of chemosensory disorders. In T. V. Getchell, R. L. Doty, L. M. Bartoshuk, & J. B. Snow, Jr. (Eds.), *Smell and taste in health and disease* (pp. 445–462). New York: Raven.

Solms, J., & Wyler, R. (1979). Taste components of potatoes. In J. C. Boudreau (Ed.), *Food taste chemistry* (ACS Symposium 115, pp. 175–184). Washington, DC: American Chemical Society.

Stone, H., & Sidel, J. L. (1993). *Sensory evaluation practices* (2nd ed.). San Diego: Academic Press.

Trant, A. S., & Pangborn, R. M. (1983). Discrimination, intensity, and hedonic responses to color, aroma, viscosity, and sweetness of beverages. *Lebensmittel-wissenschaft & Technologie, 16,* 147–152.

Uttal, W. R. (1969). Emerging principles of sensory coding. *Perspectives in Biology and Medicine, 12,* 344–368.

van Buuren, S. (1992). Analyzing time-intensity responses in sensory evaluation. *Food Technology, 42,* 101–104.

Van der Wel, H. (1972). Thaumatin, the sweet-tasting protein from *Thaumatococcus daniellii* Benth. In D. Schneider (Ed.), *ISOT IV* (pp. 226–233). Stuttgart: Wissenschaftliche Verlagsgesellschaft.

Van der Wel, H. (1993). Some thoughts about thaumatin binding. In M. Mathlouthi, J. A. Kanter, and G. G. Birch (Eds.), *Sweet-taste chemoreception* (pp. 365–372). London: Elsevier.

Wang, Y., Erickson, R. P., & Simon, S. A. (1993). Selectivity of lingual nerve fibers to chemical stimuli. *Journal of General Physiology, 101,* 843–866.

Watson, C. S. (1973). Psychophysics. In B. B. Wolman (Ed.), *Handbook of general psychology* (pp. 275–306). Englewood Cliffs, NJ: Prentice-Hall.

Woodworth, R. S., & Schlosberg, H. (1954). *Experimental psychology* (revised ed.). New York: Holt.

Wyszecki, G. (1986). Color appearance. In K. R. Boff, L. Kaufman, & J. P. Thomas (Eds.), *Handbook of perception and human performance: Vol. I, Sensory processes and perception* (pp. 9-1–9-57). New York: Wiley.

Yamaguchi, S. (1987). Fundamental properties of umami in human taste sensation. In Y. Kawamura & M. R. Kare (Eds.), *Umami: A basic taste* (pp. 41–73). New York: Dekker.

Yamaguchi, S. (1991). Basic properties of umami and effects in humans. *Physiology & Behavior, 49,* 833–841.

Yamamoto, T., & Kawamura, Y. (1981). Gustatory reaction time in human adults. *Physiology & Behavior, 26,* 715–719.

Yamamoto, T., & Kawamura, Y. (1984). Gustatory reaction time to various salt solutions in human adults. *Physiology & Behavior, 32,* 49–53.

Yamashita, H., Theerasilp, S., Aiuchi, T., Nakaya, K., Nakamura, Y., & Kurihara, Y. (1990). Purification and complete amino acid sequence of a new type of sweet protein with taste-modifying activity, curculin. *The Journal of Biological Chemistry, 265,* 15770–15775.

Yokomukai, Y., Cowart, B. J., & Beauchamp, G. K. (1993). Individual differences in sensitivity to bitter-tasting substances. *Chemical Senses, 18,* 669–681.

Zapsalis, C., & Beck, R. A. (1985). *Food chemistry and nutritional biochemistry.* New York: Wiley.

Zwillinger, S. A., & Halpern, B. P. (1991). Time-quality tracking of monosodium glutamate, sodium saccharin, and a citric acid-saccharin mixture. *Physiology & Behavior, 49,* 855–862.

CHAPTER 4

Olfactory Psychophysics

Harry T. Lawless

I. INTRODUCTION

Much has been learned in the past 25 years about the capacity of humans to recognize odors. However, a casual perusal of the literature on olfactory perception is unsatisfying due to two apparent imbalances in published research. First, a major route for bringing molecules to the olfactory epithelium, especially from foods in the mouth, is the so-called retronasal pathway. In this mechanism, volatile substances from foods diffuse, are pumped during mouth movements or pass during exhalation from the nasopharynx into the rear of the nasal passages and thence to the vicinity of the olfactory epithelium high in the nose. This is the opposite direction from inspired air or normal sniffing. There is relatively little information specific to this type of smelling or how it may differ from smelling via the inspiratory or orthonasal route. Although some speculation has been directed at the dual nature of the olfactory sense (Rozin, 1982), there has been limited study of basic questions concerning airflow patterns (Mozell, Kent, Scherer, Hornung, & Murphy, 1991), spatial patterns in stimulus absorption, or changes in qualitative perception of foods as a function of sniffing versus eating. Anecdotes persist about how one's pipe tobacco or cigar smells different when smoked by someone else. This difference probably has a lot to do with selective adaptation to different components when smoking versus smelling. However, the general questions remain largely unaddressed. This seems unfortunate when we consider how much of food flavor is determined by

Tasting and Smelling

factory impressions, and a reasonable argument can be made that (retronasal) olfactory flavor perception from foods is a more frequent part of everyday life than purposeful orthonasal sniffing of odors.

The second imbalance in the literature can be seen in the predominance of intensity-related psychophysical studies relative to odor quality perception studies. For example, sophisticated quantitative models have been compared for the description and prediction of odor mixture intensities (Sühnel, 1993), whereas almost no information exists on principles of qualitative blending of different odorants to form mixture percepts. This arises in part from the fact that quantitative psychophysical measurement of intensity of threshold is thoroughly developed but the study of qualitative perception relies principally on indirect measures such as similarity scaling (Schiffman, 1974) or odor naming.

A third important consideration when approaching the literature on smell is to recognize that any type of molecule introduced into the nasal passages is likely to have more effect than simple olfactory stimulation or imparting only odor sensations in the pure sense. Volatile substances are capable of stimulating trigeminal nerve endings as well as olfactory receptors. Although the trigeminal nerve endings are responsible primarily for sensations of irritation, the qualitative variation in this additional nasal chemical sense remains unclear. It is possible that a variety of pungent sensations are induced by different chemicals, as one would surmise from introspection on the nature of various spices (Cliff & Heymann, 1992). A growing literature has emerged on nasal trigeminal sensation (e.g., Cain & Murphy, 1980; Cometto-Muñiz & Hernandez, 1990). This chapter is limited primarily to the classical olfactory sensations; however, a review of trigeminal psychophysics can be found in Green and Lawless (1991). There has also been a growing interest in the demonstrations that humans possess a vomeronasal organ and the possible role it may play in nasal chemoreception and sexual behavior (Taylor, 1994).

A. Requirements for a Comprehensive Theory of Odor Perception

Many issues that require consideration in any theory of odor perception have received little attention from the psychological community. First, the number of differentiated odor types or odor categories is quite large. This fact works against attempts to reduce the number of odor qualities to a few dimensions, attributes, or superordinate categories. A few simple dimensions or categories will seem abstract and the whole system will appear oversimplified. Unfortunately, the cognitive load required for a comprehensive appreciation or a global scheme for odor qualities is likely to be quite heavy. Such systems will thus seem overly detailed and even arcane to the newcomer to olfactory science, and there is a natural reluctance to try to grasp the immensity of the qualitative range of odor types. This discourages some workers from even attempting to present a unified view of odor quality space (Murphy, 1987). It may appear to outsiders that odor quality schemes are domain specific or part of the arcane wisdom of various application experts, such as perfumers, when in fact there is more agreement among many systems than the newcomer would imagine (Brud, 1986; Civille & Lawless, 1986).

Second, a theory of odor perception must address the dynamic changes that occur in olfaction as a function of experience. Although difficult to demonstrate in the laboratory (Engen, 1982), the ability of perfumers and flavorists to process and manipulate their olfactory perceptions is widely believed, and the economic success of using expert noses and trained sensory evaluation panels in industry provides at least some substantiation to the claim that odor perception sharpens with practice. In fact, the skepticism of some olfactory researchers toward this expertise is surprising in light of the wide literature on perceptual learning in other modalities. The abilities of winetasters were an early interest of Gibson and Gibson, as evidenced in the following quote: "The gentleman who is discriminating about his wine shows a high specificity of perception, while the crude fellow who is not shows a low specificity. . . . If he is a genuine connoisseur and not a fake . . . he can consistently apply nouns to the different fluids of a class and he can apply adjectives to the differences between the fluids" (Gibson & Gibson, 1995, p. 35). A good deal of this enhanced discriminative ability is olfactory, because both the (sniffed) aroma and the retronasally perceived volatile flavors are sensed through smell. The question of enhanced abilities among wine experts has seen a resurgence in current interest from cognitive and linguistic scientists (Lehrer, 1983; Solomon, 1990).

The principle that expertise shapes odor quality perception has several important implications. First, any search for predetermined anatomically based "odor primaries" may be doomed to failure at the behavioral level. If olfaction functions as a synthetic sense for pattern recognition of chemical mixtures, then the guiding principle of perception organization may be that the behaving organism learns to categorize on the basis of combinations of neural patterns, resulting from naturally occurring mixtures of chemicals. Furthermore, each type of odorant molecule may stimulate a variety of (chemically) tuned receptors. So both the nature of odor stimuli as mixtures and the fact that organic molecules have more than one functional part and that each part has partial affinity for a number of receptors lead to a pattern of response. Although the study of single chemicals with highly specific binding affinity may lead to important insights into the physiological processes of olfaction, the behavioral relevance of identifying such psychological primaries is far from clear.

A second implication of the experimental effects is that the domain of expertise may shape one's categorical structure and quality perception. It may not be surprising that perfumers differ somewhat from winetasters in the distinctions they make or in the superordinate categories they employ. Some of the perfumer's categories seem to contain strange bedfellows, e.g., "balsamic," which can subsume some rich woody notes and also sweet vanillic smells. This connection seems odd because wood is rarely associated with food. However, it is understandable in terms of the perfumer learning categories based on the raw materials employed, which include some wood saps that are chemically complex. It is also known that vanilla-related notes may be infused into wine from the process of wood aging.

A third requirement for a theory of odor perception is that the nature of emotional responses to odors must be considered. Odors shape a subtle attitude, mood, or emotional tone in many experiences. This can be seen as preparatory to other behaviors, including some quite basic, such as sex, feeding, aggression, or flight. A theory that addresses olfaction only from the perspective of cognitive information processing will miss this important aspect. On the other hand, much has been made over the anatomical connection between olfactory organs and the limbic system, which of course is instrumental in emotional response and memory. Such statements often have the status of gratuitous physiologizing or truisms. To state that odors are involved in emotions is like saying that the eyes are involved in vision or the legs in running. True, but not altogether insightful. There is an unmet need for more specific and detailed theory regarding the interplay of smell and mood (Lawless, 1991).

To these three principles of olfactory perception we can add three conjectures. The first is that traditional feature models of pattern recognition may not apply to odors. Odors may be recognized as whole patterns, and the features that exist on a chemical or physiological level may not be accessible to consciousness. Our points of reference are learned by associating objects in the real world to the odors they emanate. As noted above, real world objects give off chemical mixtures with hundreds of suprathreshold components. These complex patterns are somehow synthesized into singular experiences, which, if encountered often enough, take on a well-remembered name (coffee), which can then act as a similarity reference when new, uncategorized smells are encountered ("This dark stout has roasted coffee-like aromas"). However, specifying the features of such a stimulus is difficult, if not impossible in physical terms.

A second conjecture is that not only feature models but theories of categorization based on prototypes (Lakoff, 1986) or family resemblance notions (Rosch & Mervis, 1975) may also be difficult to test. Researchers using odors invariably experience some frustration with the existing literature on human categorization. Some theories may eschew the restrictive similarity modes that count up feature overlap, but researchers use stimuli with easily manipulated dimensions or attributes (usually visual stimuli). Such approaches are irrelevant when the dimensions, attributes, or features are unknown or holistic gestaltlike processing is the rule. The olfactory scientist may need to bring new approaches to the study of categorization, rather than force-fitting odor theory to categorization in other perceptual modalities.

A third conjecture is that hierarchical structure may be critical to understanding odor categorization. Some odors are related to one another not only by perceptual similarity, but by membership in a common class. Orange and lemon odors are both citruslike. The concept of citruslike may have enormous cognitive and practical utility, especially to olfactory engineers (flavorists, perfumers), who analyze and create fragrances and flavors for popular consumption. One approach to understanding olfactory expertise is that the hierarchical structure may differentiate in both directions—from some basic experience-based level to capture additional distinctions and nuances and also to establish superordinate categories for facilitating cognitive economy. In this regard, olfaction may resemble other knowledge domains. However, this is likely to be a bumpy road at times. Odor categories may be fuzzy, with multiple or probabilistic membership of items in groups, especially to laypersons.

Overlaid on all these problems and concerns is the possibility that individuals may have strong differences in their physiological equipment and may thus move about in different perceptual worlds. This is due to the phenomenon of specific anosmia (Section II,B), a partial smell blindness to a family of closely related chemical compounds (Amoore, 1971). Even among compounds that span a normal range of sensitivities, i.e., with no obvious smell blindness, there may be stable differences among individuals (Polak, Fombon, Tilquin, & Punter, 1989). If such differentiation exists, it is a wonder that we can converse about smells and flavors at all. And yet we have some common notions about what an apple smells like, and seem to transcend the individual differences in equipment to have sensible discussions about apple smells. How this linguistic and cultural consistency is achieved should be a focus for researchers in olfaction in general and anosmia in particular.

II. DETECTION AND THRESHOLDS

A. Measurement of Thresholds

The most common index of olfactory acuity in humans has historically been the detection threshold. This has strong appeal because it measures a feature of perception and performance in physical units of concentration, and thus appears less subjective than many other psychophysical measures. The persistence of this measure is to some degree surprising, because signal detection theory calls into question some of the assumptions of threshold models, in particular the way in which they are normally conceptualized by laypersons as all-or-none boundaries. In a system that is characterized by variability—variability within individuals, across individuals, and within the stimulus—it seems dangerous to put too much stock in a measure that can be interpreted as a single concentration value, above which there are sensations and below which there are not. Even when the threshold is properly conceived as an arbitrary point on a psychometric function plotting some measure of detection probability against concentration, the functions for individuals will show reversals or "notches" (Doty, 1991a), attesting to the complications present in what at first appears a straightforward assessment. An alternative to the concept of threshold is given in signal detection theory, which recognizes that the perceivability of a stimulus emerges gradually as the stimulus level is raised above the background, and that the noise inherent in the neural system and in the atmospheric background will necessarily lead to some incidence of responding to background noise alone. A brief discussion of signal detection theory can be found in Doty (1991b) and a more thorough treatment, including the arguments against threshold models, is in Mac-Millan and Creelman (1991).

A variety of techniques have been used to measure thresholds (Doty, 1991a). Perhaps the most common is an ascending concentration series combined with a forced-choice test (ASTM, 1991). The forced-choice test answers some of the concerns of signal detection theory, i.e., that without the use of blank trials, subjects may set their own criteria for response or nonresponse, and thus their measured threshold will reflect not only their sensitivity to the stimulus compound, but their response bias as well. Obviously, changing the number of stimulus and blank presentations at each concentration level will render the task harder or easier, and readers of the odor threshold literature need to be cautioned that different methods yield measured thresholds at different levels of discriminability and that more difficult tests with lower levels of being correct by chance/guessing will necessarily yield higher estimates of threshold (e.g., Antinone, Lawless, Ledford, & Johnston, 1994). Ennis (1993) gives tables of probability of correct responses versus signal detection measures of discriminability for several common forced-choice tasks.

One method that is popular in clinical testing, due to its efficiency, is the staircase procedure, whereby concentrations ascend after incorrect forced choices, and descend after correct choices (Doty, 1991b). After several reversals, the threshold is taken as the average of the reversal points, usually omitting early run data, which are often unstable and higher than later points of discrimination, after the subject has "warmed up." Because such "adaptive methods" tend to bracket the critical area where detectability changes, they are thought to involve fewer trials to estimate threshold and thus are more suitable for clinical patient testing where the task demands are a major concern.

The detection threshold can be thought of as a specific case of discrimination, one in which the comparison is made to a blank, background, or otherwise neutral condition. This idea has been used explicitly in the development of cost-efficient test procedures to assess the detectability of flavors in various background media such as foods and beverages. One such technique uses ratings for degree of difference from a control, and assesses the points at which concentrations are statistically differentiated from the control (Lundahl, Lukes, McDaniel, & Henderson, 1986). Unfortunately, this approach has the property that group thresholds will be lower than individual thresholds because the number of observations is larger (a nonsensical situation). To remedy this Marin, Barnard, Darlington, and Acree (1991) kept the rated difference task, but suggested other criteria for determining the threshold level. They noted that in the dose–response curve there would be an inflection point for suddenly increasing ratings as the concentration level emerged to become perceivable in the background medium. Various mathematical criteria could be brought to bear on determining this point (analogous to a peak in the second derivative) and the problem of statistical sample size could then be avoided. The notion of this inflection point is potentially important for determining a psychophysical model for scaling data, as discussed below.

Another major use of threshold measures is in analytical flavor component identification. The focus there is not so much on human performance, but on the relative potency of different olfactory stimuli. Potency is usually defined as the inverse of the threshold, and flavor chemists sometimes refer to odor units as the concentration of a particular substance in a food divided by its threshold. Although such multiples-of-threshold ignore the possibility that sensation intensity might grow with concentration at different rates for different compounds (Frijters, 1978), the concept remains popular with flavorists, who must specify which compounds in a complex natural product are contributing to its smell and which are unlikely to be major players. However, even flavor chemists, who use thresholds and odor units regularly, will admit that the concentration multiple idea will not reflect subjective intensity as concentration grows. For example, Meilgaard warns that "the use of thresholds requires much caution and is not applicable above three to six odor units" (Meilgaard, Civille, & Carr, 1991, p. 124).

Nonetheless, thresholds demonstrate that the olfactory system is exquisitely sensitive to some odor compounds and remarkably passive in the presence of others. Examples of high-potency compounds would include pepper pyrazines and various mercaptans used to odorize fuel gas, which are detected in the range of parts per billion in air or less. On the other hand, the sense of smell is relatively insensitive to simple hydrocarbons and small unreactive molecules such as alcohols (Geldard, 1972). Compilations of odor threshold values have been published and include information on test method and medium of dilution, both of which affect the obtained values (ASTM, 1978).

One additional difficulty in assessing human sensitivity arises. Thresholds are variable within individuals over time (Punter, 1983; Rabin & Cain, 1986; Stevens, Cain, & Burke, 1988). Practice effects are well known in odor thresholds, and may transfer between odorants (Engen, 1960; Rabin & Cain, 1986). Overlaid on such systematic effects is a sometimes large unexplained variation for individual odors. This is not entirely a function of stimulus control or "noise at the nose," as may be the situation in variability of differential sensitivity (Cain, 1977a). The median intersession correlation (a measure of retest reliability) for a single odor was 0.61 in the Rabin and Cain study using simple sniffing bottles, whereas the Punter (1983) study found a median retest correlation of 0.4 using a sophisticated olfactometer. Rabin and Cain commented that this reliability was representative of their sample of normal young adults with a limited range of sensitivities. They also noted that "a test of session to session variation would be more meaningful if performed on participants with a wider range of sensitivities, such as clinical patients" (Rabin & Cain, 1986, p. 284).

A landmark study in testing within-individual variation in smell sensitivity was that of Stevens et al. (1988). They tested three subjects over an extended period of time and found massive fluctuations in the sensitivity of these subjects to three odorants (butanol, pyridine, and PEMEC, a rose-smelling alcohol). In fact, the distribution of individual thresholds over 20 measurements was similar to the variation observed in the population as a whole when many persons were tested a single time. Using techniques with good day-to-day reliability, drift in sensitivity to diacetyl, carvone, and cineole has recently been documented over a period of months (Lawless, Corrigan, Thomas, & Johnston, 1995; Antinone et al., 1994). These kinds of results question whether single estimates of thresholds for individual characterize any meaningful or stable characteristics of that person (Stevens & Dadarwala, 1993). If they do not, then the implications are that short-term testing may ultimately not be very useful for clinical purposes or assessments of individual differences in studies of aging or studies of other individual characteristics believed to correlate with smell.

B. Individual Differences and Anosmia

One of the truisms of olfactory perception is that there are wide differences among individuals in sensitivity, discriminative ability, and capacity to recognize and identify odors. How much these differences are a function of culture, experience, or genetics is open to question (Wysocki, Pierce, & Gilbert, 1991). One striking example of individual differences is the inability to smell a single odor compound among persons with otherwise normal olfactory acuity. This insensitivity often extends to a small group of compounds with related structures and has been termed specific anosmia (Amoore, Venstrom, & Davis, 1968), a concept originally credited to Guillot and popularized by Amoore (1971). Amoore showed that a systematic study of thresholds for related compounds such as small aliphatic acids with straight and branched chains could yield one compound that showed the greatest variation among sensitive and insensitive individuals. In the earliest study this extreme dimorphism was to isovaleric acid, which was termed *the isovaleric acid* or *sweaty odor anosmia*. In another study, a bimodal threshold distribution for isobutryaldehyde was found with modes separated by 8 or 9 binary steps. When tested with isobutyl alcohol, the modes of the two groups were separated by only 4 binary steps, and when tested with isobutyl isobutyrate, the groups were virtually indistinguishable (Amoore, Forrester, & Pelosi, 1976a). Amoore thus termed this effect the specific anosmia for isobutyraldehyde. Anosmic individuals are generally classified as those with thresholds more than two standard deviations below the population mean. Statistical distributions typically show a large mode for the bulk of the population and a smaller high threshold mode for the insensitive group (Amoore et al., 1968; Amoore & Steinle, 1991).

Amoore (1971, 1977) felt that these deficits were potentially illuminating as to the ways in which odor molecules were sensed and coded, and suggested that the specific anosmias could provide a criterion for determining what were primary odors. To the extent that odor receptors might be proteins with somewhat specifically tailored binding characteristics, this idea has intuitive appeal. On the basis of the known and suspected specific anosmias, the number of primary odors would

TABLE 1   Frequency of Anosmias to Odors[a]

Odorant	Individuals anosmic (%)	Odor type
Isovaleric acid	3	Sweaty
Trimethyl amine	6	Fishy
l-Carvone	8	Minty
Pentadecalactone	12	Musky
1-Pyrroline	16	Spermous
1,8-Cineole	33	Camphor
Isobutyraldehyde	36	Malty
Androstenone	47	Urine

[a] Data from Amoore and Steinle (1991).

then be between 25 and 50. Others have questioned this idea (Chastrette & Zak-arya, 1991), on the basis of the apparent broad tuning of chemoreceptors to various ligands and the fact that small changes in molecular characteristics will sometimes lead to major changes in perceived odor quality.

Beginning with the first reports of anosmia to isovaleric acid in 1968, a succession of papers followed rapidly with good sampling of individuals (over 50 and sometimes several hundred) and often an examination of a range of chemical compounds with similar structures. The methodology generally followed the technique of Amoore using two target odors embedded in three blanks at each ascending concentration step. Examples (see also Table 1) include the anosmia to the musky odor of pentadecalactone, also known as exaltolide (Whissell-Buechy & Amoore, 1973), the spermous odor of pyrroline (Amoore, Forrester, & Buttery, 1975), the fishy odor of trimethyl amine (Amoore & Forrester, 1976), the malty odor of isobutyraldehyde (Amoore et al., 1976a), the urinelike odor of androstenone (Amoore, Pelosi, & Forrester, 1976b), the minty odor of carvone (Pelosi & Viti, 1978), and the camphor odor of cineole (Pelosi & Pisanelli, 1981).

Wysocki and Beauchamp (1991) caution that due to factors such as differences in test methods, estimates of the frequency of some anosmias, such as that to hydrogen cyanide, vary widely (from 0 to 53% for females). Also, some of the anosmias seem more dramatic or robust than others. For example, Beets (1982) warned that the reported anosmia to carvone described only a deficit of a few binary dilutions. This might be within the retest reliability of some threshold measurements, calling into question the consistency and importance of the effect. Reports of anosmia-like insensitivities come up in other literature in some cases. For example Brennand, Ha, and Lindsay (1989), in measuring thresholds of some branched-chain acids due to their occurrence in dairy product flavors, mentioned that some subjects seemed notably insensitive to one or two compounds (and seemingly normal to others), an effect reminiscent of the isovaleric acid anosmia.

The potential genetic determination of specific anosmia was suggested in a study of inheritance of insensitivity to the musk compound, pentadecalactone (Whissell-Buechy & Amoore, 1973). Patterns of inheritance were consistent with a simple autosomal recessive trait determining nonsmelling, although a polygenic mechanism could not be completely ruled out. More conclusive evidence for genetic determination was found in a study of identical and fraternal twins and their thresholds for androstenone (Wysocki & Beauchamp, 1988). Androstenone, a steroid component present in the fat of sexually mature boars, is sensed by about 35% of people as having a foul, urinous, sweaty-type odor. A small group of 15% finds it to have a mild pleasant odor and about 50% of people can smell nothing at all. Because the presence of boar taint can dictate the condemnation of carcasses by inspection personnel, the odor is one of commercial consequence. The problem of individual variation in the ability to detect steroid compounds has long been recognized by food scientists (Thompson & Pearson, 1977). In the genetic study, thresholds for 17 pairs of identical twins were highly correlated ($r = .95$) and pairs were 100% concordant in their classification into sensitive and insensitive groups. Fraternal twins were only slightly correlated ($r = .22$) and were concordant in 13 of 21 cases. Thresholds for pyridine, included for comparison and as a control for general anosmia, were uncorrelated in both groups. This finding suggested the simple theory that specific anosmia might be under tight genetic control. Additional studies of family inheritance suggested a sex-linked trait with some X-chromosome inactivation (Wysocki & Beauchamp, 1991).

This impression was modified by later findings. Anecdotal reports of laboratory personnel working with androstenone indicated that an insensitive individual might not remain that way given enough exposure to the compound. An experimental test of exposure showed that daily smelling for three periods of 3 minutes for 6 weeks was sufficient to cause a substantial decrease in threshold (increase in sensitivity) in 10 of 20 anosmia subjects (Wysocki, Dorries, & Beauchamp, 1989). Little or no changes were seen among a control group retested at the same intervals. This suggested possible induction of receptors that were previously in small quantities or unexpressed, or induction of an enzyme system involved in androstenone reception. Massed exposure for three 90-minute sessions did not lead to the change noted in the 6-week experiment (Wysocki & Beauchamp, 1991). Developmental factors may also play a part. Dorries, Schmidt, Beauchamp, and Wysocki (1989) found that prepubertal children were largely able to smell androstenone but that the incidence of anosmia among males increases dramatically around puberty. In contrast, mean thresholds for females decrease over adolescence and a larger proportion remain smell sensitive. Threshold changes were closely paralleled by hedonic evaluation, with an increase in neutral or positive ratings accompanying the onset of anosmia.

The robust effect found in the original twin study suggests that the androstenone anosmia is largely stable. This conclusion must be modified somewhat to account for the effects of exposure and development, as noted above. Other investigators have found stable individual differences in the sensitivity to enantiomeric pairs of ionone compounds (Polak et al., 1989). In that study, some subjects were more sensitive to one member of the pair and other subjects showed the opposite pattern.

Nine retested subjects retained stable ratios of sensitivity to the pair over several months.

The stability of other reported anosmias is open to question. As previously noted, thresholds are variable within individuals. The wide variation noted by Stevens et al. (1988) should raise the question of whether specific anosmia may be a stable condition or a transient passage of otherwise normal individuals into a temporary high threshold state (Lawless et al., 1995). In our laboratory, specific anosmias to cineole, carvone, and diacetyl have been studied due to their potential importance in food flavor perception. Cineole and carvone are common terpene aroma compounds; diacetyl occurs in many dairy products and in beer as a function of microbial fermentations. Our experiments showed a disturbing pattern of individual sensitivity fluctuation for all three compounds, which rendered classification as anosmia somewhat moot in terms of predicting stable trends in food flavor perception (Lawless et al., 1995). Specifically, the following effects were observed: First, moderate correlations of threshold with suprathreshold responsiveness were found when both threshold and suprathreshold measurements were taken in the same session. However, suprathreshold scaling in a mixture study conducted 2 months later showed no difference in ratings between groups previously classified as anosmic versus normal. Second, wide individual variations were noted for carvone sensitivity. However, when thresholds in high- and low-sensitivity subgroups were retested after 3 months, differences between the groups were virtually gone. Retesting thresholds for cineole after 18 months showed similar "smearing" of sensitive and insensitive groups due to individual shifting. Third, thresholds for diacetyl also show a pattern consistent with a high-threshold subgroup (Lawless, Antinone, Ledford, & Johnston, 1994). Furthermore, discrimination of diacetyl in cottage cheese is highly variable. Classification of two groups as good versus poor discriminators was done on the basis of triangle and 3AFC tests to a range of diacetyl concentrations added to cottage cheese. Retesting good versus poor discriminators 1 month later showed no pattern of consistent difference between the two groups (Antinone et al., 1994).

Taken together, these results suggest that the individual variation noted by Stevens et al. (1988) may also be present for compounds to which there are specific anosmias. If so, at least some specific anosmias may represent merely transient conditions. This would imply that specific anosmia is not linked to the absence of genes directing the development of "primary" receptors, but is rather a function of some other, as yet undetermined, cause of olfactory sensitivity changes, e.g., changes in nasal airflow or in the perireceptor environment. The question also remains as to whether these shifts are general effects or whether they are specific to individual compounds or families of closely related compounds. The results of the study by Stevens et al. did not mention synchronous shifts for their three compounds, and more data on this matter would be helpful.

Another unresolved issue in the study of individual differences concerns patterns of intercorrelation among compounds. If specific anosmias are primaries, should they be unrelated? To the extent they are conceptualized as separately functioning channels, there would be no a priori prediction of any correlation across individuals. However, the extent to which the genes controlling the expression of these effects are correlated might be an underlying factor. In some of the tabulations by Amoore (1971, 1977) there was an attempt to group reported anosmias for related compounds (e.g., the goaty-smelling acids), but there has been little actual testing of co-occurrence. Brown, Maclean, and Robinette (1968) and Punter (1983) argued that there was little association among odorants in individual sensitivity other than that explained by a general sensitivity factor. Polak et al. (1989) found little association between sensitivities to the two isomers of ionone. Beets (1982) reviewed the large body of literature on structure–activity relationships among the musks and concluded that there was a single primary odor ("modality" in his terms), although he noted some experimental reports in disagreement. In general, the anosmias to musk compounds are only a few binary steps and the picture is far from clear. Beets also suggested an association among the insensitivities to the steroids and various cyclic ketones. This was experimentally tested by O'Connell, Stevens, and co-workers, who documented an association between sensitivity to androstenone and pememone (O'Connell, Stevens, Akers, Coppola, & Grant, 1989; Stevens & O'Connell, 1991). Factor analysis in the study of O'Connell et al. (1989) also showed patterns of inverse relationships of responsiveness to the urinous odors and to phenyl ethyl alcohol (having a rose odor), and also inverse relationships between pemenone and pepper pyrazine. Another report questioned the possible correlation between Galaxolide, a commercially important musk fragrance compound and androstenone (Baydar, Petrzilka, & Schott, 1993). Although there was a statistically higher incidence of persons anosmic to both compounds than would be expected by chance, many of the persons studied were anosmic to one, but not the other, compound. Furthermore, the association was stronger among women than men. This raises many unanswered questions regarding the mechanisms of association of different odor sensitivities and their possible sex-linked modulation. Finally, given that many of the anosmias to food-related compounds are to small carbonyl compounds of similar chain length, the possibility of associations deserves further attention.

III. INTENSITY RELATIONSHIPS

A. Discrimination Issues

Early attempts at assessing the discriminative powers of the sense of smell borrowed from the information sciences arising in the 1950s. In a typical experiment, several presumably reproducible stimuli would be set up and the subject would be provided with well-rehearsed labels for the stimuli. On subsequent trials, the stimuli would be presented and the subject asked to label them. Identification and misidentification would be analyzed by the familiar Shannon–Weaver equation for information transmission and a channel capacity would be estimated. From such studies, it was estimated that the sense of smell is not very good at giving information on intensity on an absolute basis, with an estimated channel capacity of only 1.5 bits (Engen & Pfaffmann, 1959). By *absolute basis,* it is meant that a unique label must be assigned to a single stimulus, without direct comparison to any other available stimuli. This suggests that without any additional information or comparisons, the human nose identifies only about three levels of smells with any accuracy, and attempts at higher levels of performance are doomed to failure.

There is substantial corroborating evidence for the generalization that humans, without extensive training or practice, will not discriminate well among odor intensities. First, it is probably no coincidence that applied methods for sensory evaluation of products usually begin with a simple three- or four-category scale for intensity ratings. As an example, the A. D. Little Flavor Profile Method uses a "none–weak–moderate–strong" scale (0 to 3) (Caul, 1957) and the standard methods for judging dairy products use a "slight–moderate–pronounced" scale (Bodyfelt, Tobias, & Trout, 1988). Panels with training learn to use additional scale divisions, but the naive nose and brain clearly are saddled by some limitations in the olfactory modality. In quantitative descriptive analysis of foods, all sensory aspects of a product are evaluated, and it is a common finding in the univariate analyses of variance that the aromas and volatile flavors will show the lowest F-ratios among a group of products. These aspects are simply noisier and harder to evaluate consistently than the visual and gustatory qualities of foods. Finally, it is a general rule of thumb among product formulators that flavor ingredient levels must change by large factors (50–200%) in order to have sensory changes that will be noticed by consumers.

It can be argued that this limitation on intensity identification is another reflection of the poor discriminative power of the sense of smell relative to concentration and perhaps the lower power function exponents observed in magnitude scaling as compared with other modalities. In reviewing the historical literature on differential sensitivity, Cain (1977a, 1977b) reports the Weber fraction for smell to fall in the range of about 25 to 45% for many odorants. This is about three times the change needed to discriminate auditory or visual stimuli. He went on to examine how much of the confusibility of odorant concentrations could be due to undetected variation in the physical stimulus. Odor discrimination studies are often done with sniff bottles whereby the subject samples the vapor phase or headspace over a liquid of known concentration. Cain sampled the vapor phase with a gas chromatograph and determined that the headspace could vary in concentration quite a bit from the nominal concentration present in the liquid phase. This unwanted variation was highly correlated with discrimination performance, with stimulus variation accounting for 75% of the variance in discrimination. This strongly suggests that historical estimates of difference thresholds are too high, due to difficulties in stimulus control and delivery. Use of an olfactometer, instead of sniff bottles, dropped the value of the Weber fraction to 4%.

Although these observations are psychophysically and physiologically important, their relevance to everyday life may be limited. Smells in the real world are rarely delivered by olfactometers, and the airstreams and concentration flux normally encountered are probably great sources of variation. From an ecological perspective, the utility of concentration differentiation for humans may be limited. This would obviously be not so for a species that hunted or homed via concentration gradients in chemical stimuli.

B. Psychophysical Functions

Because the threshold is only one point on a dose–response curve, it may seem surprising that so much attention has been paid to it. This situation is partly remedied by studies of sensation intensity above threshold using various scaling methods. This provides a more complete picture of the dynamic response of the olfactory sense with increasing stimulus concentrations. Because odor stimulation is necessarily accompanied by a stimulus in flux, that is, molecules in a moving air stream, one can consider flow rates or measures of flux (molecules exposed to receptors per unit time) as determinants of physical stimulus potency as well as concentration. Some changes in sensation appear to be a function of the number of molecules reaching the receptors (Mozell et al., 1991), but other investigators have found increases in sensation as a function of flow rate per se (Rehn, 1978). A review of this problem can be found in Mozell et al. (1991), who give important warnings about the possible confounding of flow rate measures with total number of molecules, dwell time, and other variables.

A variety of procedures have been employed for assessing responses above threshold, including magnitude estimation, category scales, line-marking techniques, and others (Doty, 1991a; Lawless & Malone, 1986). The line-marking technique, currently popular in industry, was anticipated by a cross-modality match to distance via finger span (Ekman, Berglund, Berglund, & Lindvall, 1967). To some extent, the form of the function that fits the data depends on the technique used for scaling. Not surprisingly, bounded scales, such as category scales and line-marking techniques, yield highly curved logarithmic relationships, whereas unbounded scales, such as magnitude estimation, yield power functions, which, although negatively accelerated, are less restricted or bounded than the category/line data (Stevens & Galanter, 1957). These results have been thoroughly discussed in the psychophysical literature, including longstanding debate about which scale type is more "valid" than another (Anderson, 1970; 1977). This is mentioned here merely to caution the reader that the result from a scaling study may depend on the method. Over short stimulus ranges, the scale types yield very similar results and are equally discriminative regarding stimulus concentrations (Lawless & Malone, 1986). One alternative approach to scaling is to use perceived intensity matches, in which case the odor intensity can be expressed in concentration units of a reference odorant such as butanol (Moskowitz, Dravnieks, Cain, & Turk, 1974).

The most common function in the psychophysical literature is the power law popularized by S. S. Stevens (1962) and generally fit to data from magnitude estimation or other ratio-instruction tasks:

$$\Psi = k\Phi^{\beta} \tag{1}$$

or

$$\log \Psi = \log k + \beta \log \Phi, \tag{2}$$

where Ψ is the perceived intensity, Φ is the physical intensity (usually in concentration units), β is the characteristic exponent, and k is a proportionality constant. This relationship is often modified to include a correction for threshold, Φ_0, as follows:

$$\Psi = k(\Phi - \Phi_0)\beta. \tag{3}$$

The notable characteristic of psychophysical power functions for odors is that their exponents are generally less than 1, and often quite a bit lower than those found for other sensory modalities (Berglund, Berglund, Ekman, & Engen, 1971a). Engen (1982) summarized the literature by stating that a value around 0.6 would be representative. Of course, exponents less than 1 characterize a law of diminishing returns, wherein larger and larger increases in physical concentration are necessary to produce increases in sensation magnitude. The odor sense thus responds over a wide range of concentration levels and it makes sense to examine response as a function of the logarithm of concentration. The power function has been widely used for olfactory data (Cain & Engen, 1969) and is incorporated into some mixture models described below.

Some other mathematical functions have been fit to chemosensory data. The so-called Beidler function, a semihyperbolic relationship that is a transformation of the equation used in Michaelis–Menten enzyme kinetics, has been widely used in taste research (McBride, 1987) but received little application to olfactory data. As originally cast by enzyme biochemists, the equation takes the following form:

$$\Psi = \frac{(\Psi_{\max}\Phi)}{(k + \Phi)}, \tag{4}$$

where $\Psi_{\max}$ is the asymptotic sensory response and k is the concentration at which response is half-maximal, which is related to the binding affinity of the odorant, and Φ is once again physical concentration. This equation has several properties that make it appealing for olfactory data. First, when plotted as the log of concentration, the function will be ogival with an inflection point above threshold in the dose–response curve. This is a useful feature as noted previously in the threshold method of Marin et al. (1991).

Second, the function will approach an asymptote, producing the curvature seen in the power functions with exponents less than 1. An asymptote implies that the system will saturate, which makes sense for any chemical process involving a limited number of receptors. However, experiments that probe a functional asymptote are difficult to do. Irritation or pain may enter the sensation mix at higher stimulus levels owing to stimulation of the trigeminal system. This may then act to inhibit odor perception. Cain and Murphy have used this to explain why the psychophysical functions of at least some subjects will show downward trends in rated odor intensity at extremely high concentrations (Cain, 1976; Cain & Murphy, 1980).

C. Adaptation

A common experience is to notice the unique smell of a home when entering a house for the first time, but to slowly become unaware of it after a few minutes. This experience is so common that most people think that adaptation of a fairly complete nature is an important characteristic of olfaction (Engen, 1982). Adaptation is commonly described as a decrease in the perceived intensity of a stimulus over time under conditions of constant stimulation. Adaptation has sometimes been assessed (more laboriously) as a decrease in sensitivity or increase in threshold, and in one study was examined as the (increasing) concentration necessary to maintain constant sensation intensity (Cain, 1974).

From the results of such experiments, two generalizations can be made (summarized in Engen, 1982). First, higher concentrations take longer to adapt or seem to be more persistent. Second, when the concentration of a probe stimulus is varied following adaptation, weaker concentrations are more greatly affected than are stronger ones (Cain & Engen, 1969). To describe the time course of adaptation, a general equation proposed by Overbosch combined a decaying experiment with the power function (Overbosch, 1986):

$$\Psi = k(\Phi e^{-At/\Phi} - \Phi_0)\beta, \tag{5}$$

where A is a decay constant indicative of the rate of adaptation and t is time. A similar relationship was suggested for the time course of trigeminal irritation (Lawless, 1984).

A third general principle might be that olfaction is a sense that adapts completely, but experimental evidence on this widely held supposition is not supportive. Demonstrating that odors truly "disappear" is more difficult than meets the eye. One problem is that the decision that no sensation is present involves the criterion and response bias concerns emphasized by signal detection theory. In one study that examined detection of hydrogen sulfide, even weak odors show some higher hit rate than false alarm rate, indicative of perceivable sensation (Berglund, Berglund, Engen, & Lindvall, 1971b). This study also found higher hit rates and lower false alarm rates with higher concentrations, but that the duration of the adapting stimulus, varied from 1 to 5 minutes, had little effect on the detectability and, conversely, the degree of adaptation. This last result suggests that whatever degree of adaptation does occur, it does so in a matter of less than a minute, which is consistent with older literature on adaptation rate (Köster & de Wijk, 1991).

Another problem is maintaining control over a stimulus mixed with air that is being sampled in a cyclic manner. When subjects were breathing normally (but presumably able to sniff odors to sample them), adaptation reached a steady state that maintained about 40% of its initial intensity for propanol, butanol, and butyl acetate (Cain, 1974). de Wijk (cited in Köster and de Wijk, 1991) reported complete adaptation to even strong concentrations of odorants when the stimulus is passively injected through the nose in a constant stream, rather than being actively sniffed by the subject.

The degree to which observed changes in olfactory sensation as a function of time of exposure have been attributable to adaptation of peripheral receptor signals or more central mechanisms under conscious control is not well determined. Engen (1982) has argued that decrements in responding could easily be the result of habituation rather than sensory fatigue, and that in many situations, we stop attending to background stimuli that are constant but unimportant or uninformative (e.g., the ticking of a clock).

IV. MIXTURES

A. Interest in Mixtures; What Is a Mixture?

It is a widely held truism in the study of chemical sensation that naturally occurring stimuli are rarely single chemicals, and even when receptors are fairly specifically tuned, as in the case of insect pheromones, even then the chemical signature of the sending organism often constitutes a mixture of several compounds. The study of simple mixtures is thus seen as a step toward bridging the gap between psychophysical studies of single chemicals and the real life experiences of smelling and eating. Research has focused primarily on the simple issue of how the perceived intensity of a mixture can be related to or predicted by the perceived intensities of the components when unmixed but in concentrations equal to those present in the mixture. A second, less studied issue is how the quality of a mixture may change relative to the odor types of the components.

This is a difficult issue that has now been well formulated theoretically. Odors, like tastes, seem to span a range of effects, from a totally synthetic condition of blending to one where two components may be readily separable and perceived as separate notes within a smell. As an example of the first effect, a common fragrance material in lemon-smelling products is the compound citral, which is actually a racemic mixture of neral and geranial. However, the mixture is readily grasped as a singular artificial lemon-type odor, one that has high familiarity due to the presence of citral in various household cleaning products. As an example of a separable mixture, early experiments on two-compound mixtures used pyridine and lavender compounds, an easily separable two-note percept (e.g., Cain & Drexter, 1974). To add to the complexity of the situation, it appears possible for subjects to adopt either a holistic or analytical perspective in an odor perception task. That is, under certain task demands, they may adopt an analytical frame of approach to an odor, and break it apart into constituent notes. Such is the everyday job in applied sensory evaluation of food and consumer products (Caul, 1957; Lawless, 1996) by flavor profile panels who will provide intensity ratings for a number of individual aromatic properties of a complex food flavor. The extent to which applied sensory evaluators, trained perfumers, and the like are actually smelling multiple percepts is open to question. It is also possible that they may be evaluating the similarity of a homogeneous percept to various prototypical mental records when producing such a profile. The analogy here is like evaluating the redness versus yellowness of various shades of orange color. Although they are each perceived as a singular color, they are able to be described according to the familiar points of reference among the focal colors. Thus behavior may appear analytical, due to task demands, but perception remains synthetic.

A final problem in this area is that individual molecules may have multiple functional groups and thus interact with a range of olfactory receptors. Thus it is possible for a single chemical compound to have a percept that seems mixed. An example is the compound salicaldehyde, which smells partly like aspirin and partly like almonds. In perusing any catalog of flavor and fragrance chemicals (Bauer & Garbe, 1985), it is not uncommon to read descriptions like "woody and vanilla-like" or "citrus-fruity and slightly balsamic."

In summary, although most simple psychophysical experiments would like to employ compounds that produce a singular percept, such is unlikely to be the case for most chemicals occurring in nature. Two levels of complexity make the study of smell mixtures very difficult—mixtures may appear to be single smell percepts, and single chemicals may appear to produce multiple smell percepts. Thus any model that assumes a simple relationship between a single chemical and a single qualitative percept may be applicable to only very special cases of smell materials.

B. Intensity Issues: Inhibitory Interactions

A common result in mixtures of odors is that the intensity of the mixture is less than one would predict from the intensities of the components. Cain and Drexler (1974) outlined a set of outcomes that compared the intensity of the mixture to the sum of the intensities of the components, using language that had been found in the early literature. Masking was discussed in terms of quality masking, a change in the odor type as a function of adding a second odor, and masking of a single component through intensity reduction. The overall term for reduction in the intensity of a malodor as a function of additional smells being present was termed *counteraction*. *Partial addition* described a situation in which the mixture would be perceived as stronger than the stronger of the two components, but weaker than the sum of the components. *Compromise* describes a situation in which the mixture intensity is intermediate to the two components, and *compensation* describes a situation in which the mixture is less intense than either component.

Although these terms provide some clarification of how the words are used in the previous literature, the approach assumes that perceived intensity has been measured on an interval scale in which addition is an allowable operation. Such conditions are rarely met or demonstrated. For example, use of any bounded scale such as a category or graphic rating scales makes summation a fallacious exercise. If the two components are rated above midpoint on such scales, the mixture will necessarily show counteraction because it cannot be rated higher than the endpoint, and the sum will obviously be higher. Assessing overall intensity also omits what is often a more interesting question of how one component is reduced, as in the case of desiring to cover a malodor.

In general, counteraction is a robust phenomenon, as is mixture suppression in the sense of taste (Köster, 1969; Cain, 1975; Cain & Drexler, 1974; Laing, Panhuber, Wilcox, & Pittman, 1984; Laing & Willcox, 1983; Lawless, 1977). Not only is it common for the total intensity of the mixture to show some hypoadditivity relative to the sum of the components, but in cases in which the individual components can be analytically perceived, they are less intense in the mixture. This type of intensity masking is also asymmetric. Laing et al. (1984) showed that the more intense of two components would have a strong effect on diminishing the weaker component. Only when the components were perceived to be about equal in intensity would the masking be reciprocal and symmetric, otherwise the more intense component seems to dominate. Intensity masking also occurs between chemicals that have trigeminal and olfactory properties (Cain & Murphy, 1980; Cometto-Muñiz & Hernandez, 1990). Because some odorous chemicals also stimulate the trigeminal nerve at higher concentrations, it is possible to obtain inhibition of odor intensity from a single molecule, and a psychophysical function for odor intensity that is nonmonotonic, that is, that turns down at high levels (Cain, 1976).

Several mathematical relationships have been proposed for describing the intensity interactions in odor mixtures, and characterizing the degree of additivity or departure from additivity by one or two constants. Some of these models are developed in terms of the additivity of percepts (Berglund, Berglund, Lindvall, & Svensson, 1973; Patte & Laffort, 1979) whereas others specify interactions in terms of physical concentration relationships (Sühnel, 1993). Some models have taken into account both additivity of percepts and the psychophysical relationship. This has some classification value for what defines true synergy, as we see in the following discussion.

The first simple model provided to describe odor addition in mixtures is that of Berglund et al. (1973), which described the mixture relationship as a process of vector addition (the resultant of a parallelogram) with a characteristic angle α:

$$\Psi_{ab} = \sqrt{\Psi_{a^2} + \Psi_{b^2} + 2\Psi_a\Psi_b \cos \alpha}, \tag{6}$$

where Ψ_{ab} is the perceived intensity of the mixture and Ψ_a and Ψ_b are the perceived intensities of the components. The constant, cos α, would typically have negative values in the range of -0.2 to -0.5, corresponding to angles generally greater than 90° but less than 180°. This model provided a reasonable fit to many binary mixtures. However, it had two major shortcomings. First, it did not take into account the nature of the psychophysical function, and second, it was unable to describe cases of synergy.

The latter problem was taken into account in an alternative model, called the U model, by Patte and Laffort (1979). This model was also shown to produce slightly better fits to the data of Cain and Drexler (1974) than was possible with the vector model:

$$\Psi_{ab} = \Psi_a + \Psi_b + 2\sqrt{\Psi_a \Psi_b K}. \tag{7}$$

In this case the parameter K could take on negative values as in the case of cos α, but could take on values greater than zero to describe cases of synergy. This model still failed to reflect interactions that might be a result of the psychophysical relationships inherent for the growth of sensation with concentration increases. This is an important consideration, because any negatively accelerated psychophysical function (e.g., a power function with exponent less than 1) will show apparent subadditivity when added to itself. Whether this truly reflects any inhibitory interaction with physiological implications is certainly open to question. This notion was also employed in the taste mixture literature, where Bartoshuk and colleagues noted that taste mixtures from substances with exponents less than 1 tended to be hypoadditive (Bartoshuk, 1975; Bartoshuk & Cleveland, 1977). Frank, Mize, and Carter (1989) extended this idea and defined true synergy as cases in which the mixture interaction was greater than that which could be predicted from self-addition.

A similar approach was taken earlier by Laffort and Dravnieks (1982), who extended the U model to take into account the degree of additivity in self-mixtures, that is, inherent in the psychophysical relationship of a substance. They called this the UPL model, for U and power law, because they worked on the basis of Steven's power function. They defined a new index of interaction, γ, to describe the degree of interaction present in the psychophysical function relative to the degree of interaction present in mixtures:

$$\gamma = \frac{(1 + \cos \alpha_U)}{(1 + \cos \alpha_{UPL})}, \tag{8}$$

where cos α_U is the parameter from the angle estimated from the mixtures and cos α_{UPL} is estimated from the power law. Values of $\gamma < 1$ would indicate true partial inhibition, and potential interactions beyond the transformation inherent in the receptor mechanism, such as overlap in stimulation by the mixture components or competitive inhibition. This model was further refined by Laffort, Etcheto, Patte, and Marfaing (1989) to provide a more general equation for cos α_{UPL} where the model would potentially hold for all concentrations of component a and b:

$$\Gamma = \frac{(1 + \cos \alpha_{\text{U}})}{(1 + \cos \alpha_{\text{UPL2}})},$$ (9)

where

$$\cos \alpha_{\text{UPL2}} = \frac{(\cos \alpha_a \Psi_a + \cos \alpha_b \Psi_b)}{(\Psi_a + \Psi_b)}.$$ (10)

This model was found to generate isointensity curves similar to those observed experimentally by Köster (1969) under various values of Γ and different power function exponents.

A somewhat different approach based on isointensity curves was taken by Sühnel (1993) that sought to unify all the various models under one simpler concentration-based relationship. This approach is widely used in pharmacology and biomedical research. It examines the concentration ratios of components in a mixture, relative to the components that would produce the same result if unmixed, and defined the interaction index I as follows:

$$I = \frac{c_a}{C_a} + \frac{c_b}{C_b},$$ (11)

where c_a and c_b are the concentrations of the components in the mixture and C_a and C_b are the concentrations of the unmixed components that would produce the same results as the mixture. Values of $I < 1$ indicate synergy whereas values of $I > 1$ indicate subadditivity (antagonism, in Sühnel's terms). Various response-surface methods were recommended for visual examination of interactions. Experimental verification showed that values of I and Γ are highly correlated, although the descriptions of synergy versus antagonism will diverge somewhat as the power function exponents of the two substances diverge from equality.

C. Enhancement

As a general rule, inhibitory or suppressive interactions are observed in taste and smell mixtures at high intensity levels, and synergistic or simply additive mixture effects are seen at low intensity levels (Frank et al., 1989). For example, a recent study of thresholds found that complete additivity was the rule: three-component mixtures had thresholds at one-third the concentration values of the thresholds of the individual components (Patterson, Stevens, Cain, & Cometto-Muñuz, 1993). Reports of enhancement or synergy in odor mixtures are not common. Köster (1969) reported that of the 18 pairs he studied, no cases of enhancement were present. However, inspection of his figures shows some cases of small enhancements, usually when a low amount of one odor is added to another component that is predominant in the mixture.

Gregson (1986) recognized Köster's findings as a possible case of "Fechner's paradox," in which isointensity curves would show some points of synergy at very low levels of addition of the second stimulus. He showed another form of the paradox, in which estimates of the intensity of one component of the mixture would pass through a zone of enhancement when small amounts of a second odor were added. Gregson suggested one factor in odor mixtures that might explain this as well as some counteraction effects. Enhancement at low levels might result from an aqualitative intensity addition when the second odor is very weak. It seems reasonable that when low amounts of a second odor are added to a first, its quality might not be entirely definable, that is, the recognition threshold has not yet been surpassed. In this case the subject could misattribute added intensity from the second odor to the first. As the second odor becomes stronger, its own identity may become apparent and the normal function of counteraction takes over. This result has a possible parallel in taste research, where Halpern (1991) showed that tracked taste quality had a later onset and earlier offset than did tracked intensity. In other words, intensity tracking of the course of taste sensations may have aqualitative early and late segments in which some sensation is registered, but subjects are unsure of its identity. Gregson also pointed out that one process leading to odor counteraction might involve conversion of part of the quality of an odor to an aqualitative component.

A type of sequential enhancement occurs when mixtures are preceded by adaptation to one of the components of the mixture. Engen and Lindström (1962) examined the perception of amyl acetate–heptanal mixtures indirectly via similarity ratings and a ratio judgment task. They extracted factors correlated with the physical concentrations of each odorant. After adaptation to amyl acetate, the factor scores associated with amyl acetate level decreased and the scores associated with heptanal level increased. Conversely, adaptation to heptanal increased scores on the factor associated with amyl acetate level and decreased scores on the factor associated with heptanal. This is reminiscent of a well-documented effect in taste perception, wherein adaptation to one component of a mixture decreases the intensity of that component, but also increases the intensity of the other component, an effect that has been termed *release from mixture* suppression (Kroeze & Bartoshuk, 1985).

This release effect also appears to occur in olfaction, although the data are a bit noisier and the release effects are not nearly as complete or reliable as they are in taste mixture research. This difference may reflect the inherent problems in stimulus control in olfactory studies, or some differences in the mechanisms of adaptation and mixture inhibition. Nonetheless a general parallel does exist. Figure 1 shows a mixture experiment in which vanillin and cinnamaldehyde were presented in vapor-phase mixtures (Lawless, 1987). Counteraction is evident in the decrease in intensity of the rated vanilla odor component and the rated cinnamon odor component in the mixture. After adaptation to one component of the mixture, the odor

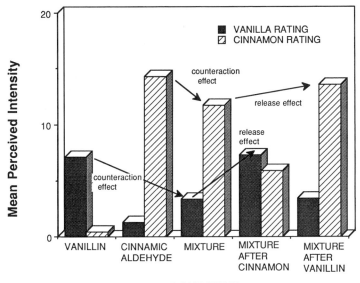

FIGURE 1   Release from mixture inhibition. In the mixture, the intensity of vanilla and cinnamon odors was less than when each odor was presented individually. After adaptation to one component, the other component increases in intensity, roughly to the level at which it was perceived when unmixed. Modified from Lawless (1987).

quality associated with the other component is increased in intensity. This occurs even though adaptation to one component was incomplete. The decrement in intensity of the adapted component is apparently sufficient to undo most of the inhibitory or counteractive effect on the second component.

D. Blending versus Component Identification

In general, odor mixtures prepared in the laboratory bear a strong resemblance in their character to the quality characteristics of the individual components. For example, Laing and Willcox (1983) showed that in binary mixtures, the odor profiles were generally similar to or predictable from the profiles of the components, although an intensity mismatch tended to favor the dominant component at strong expense to the perception of the quality of the weaker item. This would suggest that emergent qualities or deriving a wholly new odor as a function of mixing are rare. However, complex odors in nature may not be so bound by this rule. Anecdotes exist about how new patterns may arise from multicomponent mixtures in which the odor of the emergent pattern is not clearly present in any single component. For example, a mixture of 10 or so medium-chain aldehydes ($C_6 - C_{16}$) produces a smell reminiscent of wax crayons (Lawless, 1996). Furthermore, natural flavors consist of mixtures of many chemical components, no single one of which may possess the odor quality characteristic of the blend, for example, synthetic tomato aroma made from a mixture of various chemicals (Buttery, Teranishi, Flath, & Ling, 1989; Lawless, 1996). Although the odor of chocolate is a distinctive pattern, it is difficult to find a single chemical component that produces this impression—it seems to be an emergent property of a complex mixture. It is also common to find in a naturally occurring mixture many odor-active compounds that bear little or no resemblance to the product aroma, but somehow in the mixture the emergent quality is produced. For example, in the analysis of cheese aroma by gas chromatographic sniffing, the components found to be contributory had no cheese aromas in their individual characteristics, yet the aroma of cheese occurs in the mixture (Moio, Langlois, Etievant, & Addeo, 1993). Burgard and Kuznicki note that such synthesis may be the rule: "Coffee aroma is contributed to by several hundreds of compounds, a great many of which do not smell anything like coffee" (Burgard & Kuznicki, 1990, p. 65).

Mixture complexity in even simple mixtures is difficult to predict (Moskowitz & Barbe, 1977). This characteristic of smells has received little experimental attention, although there is a parallel literature in taste on estimates of singularity versus mixedness (Erickson & Covey, 1980; O'Mahony, Atassi-Sheldon, Rothman, & Murphy-Ellison, 1983). Moskowitz and Barbe (1977) found that odor complexity is not additive. Placing a highly distinctive odor (camphor) with a second component (isobutyl butyrate) results in a mixture that was judged as highly complex, although adding a third component diminishes the rated complexity. To the extent that rated complexity reflects the perceptual separability of components or lack of blendedness, this may be understandable. Addition of the third component renders the mixture too highly blended to seem complex. In other words, an new more synthetic pattern may have emerged. A similar effect was when two highly singular odors, cedarwood oil and orange oil, produced a mixture that was judged to have two components by most subjects. Conversely, a mixture of pine oil and lemon oil, which were judged as more complex (less singular) individually, was paradoxically judged less complex than the cedar/orange mixture (Lawless, 1992). Thus it would seem that more complex odors might form more blended mixture percepts, or that complexity is not necessarily conserved.

As previously suggested, odors blending to form homogeneous percepts may be commonly the rule in nature. Cooking foods results in a huge number of reaction products produced by combination and recombination of chemical constituents and by breakdown of macromolecules such as protein and starch. Maillard browning of sugars and amino acids provides common examples, as in the baking of bread, roasting of meats, or production of chocolate. The perhaps surprising result is that a mixture is produced that is a singular percept, for example, fresh-baked bread or chocolate. Even when a single chemical is largely responsible for the primary smell of a natural product (e.g., vanillin in vanilla), the informing compound is only a poor caricature of the fuller aroma of the natural essence with all its additional constituents. And yet the natural odor can still seem to be a unitary smell. This raises the question of how odor mixtures are sometimes perceived as having multiple qualities and under what conditions human observers can interpret only part of a smell in the presence of additional background. This question has implications for the sensory analysis of foods and consumer products, detection of taints and off-flavors in product quality assessment, as well as important olfactory detection tasks such as noticing odors associated with a fire or a gas leak embedded in the background odors of a home atmosphere.

Some results from mixture component identification indicate that people may be worse at this task than one first supposes. Laing and Francis (1989) showed that subjects could correctly identify at least one odor from a binary mixture of highly discriminable odors only about 40% of the time, and with additional components (up to five) performance deteriorated even further. Furthermore, subjects seemed remarkably unaware that more than two or three odors had been presented. Laska and Hudson (1992) used a number of mixtures of differing complexity (3, 6, and 12 components) and studied the discriminative ability when components were dropped. Several effects highlight poor performance in this task. In comparing 6- and 3-component mixtures, ratings of "identical" were obtained in 30% of trials. In 10% of trials, 12- and 6-component mixtures were judged to be identical. There was also a high false alarm rate: when mixtures were in fact the same, they were judged to be different in over 50% of trials.

As previously noted, trained sensory analysts employed in industrial product evaluation can presumably do this quite well. Perfumers and flavorists are also reported to act analytically in the process of duplicating a natural product or reproducing a competitor's successful flavor blend. This would suggest an important role for training and familiarization in the detection and identification of components of odor mixtures. It is a general principle of sensory evaluation that untrained consumers tend to approach a product holistically, whereas the trained panelist is more adept at analyzing and separating individual sensory attributes. This may have a parallel in development, in the sense that young children integrate some stimulus dimensions that older children can use separately (Smith and Kemler, 1977). However, Laing (1991) reported that even with training, subjects failed to identify more than three components in the four- and five-component mixture reported in the set of Laing and Francis (1989), although performance was higher than with untrained subjects.

Rabin and Cain attacked this more directly by examining discrimination of odor mixtures in which one component was less intense, and this "minor component" could vary in its familiarity and pleasantness (Rabin & Cain, 1989). Two hypotheses were tested: first, that the familiarity with the minor component would enhance the discriminability of the mixture from a stimulus containing the major component only, and second, that unpleasant minor components would also render the mixture more discriminable. This second idea was based on the observation that people seem to perform well at detecting off-odors or unwanted taints in products. Familiarity with both major and minor components enhanced discrimination, supporting the first hypothesis. Unpleasant minor components also were slightly more discriminable than were pleasant minor components. One important feature of this study was that different odorants were chosen for each subject based on their own evaluation of familiarity and pleasantness. This degree of "tailoring" is necessary in such a study and highlights the idiosyncracy of individual odor experience.

This study also examined false alarm rates in discrimination in a small number of trials when the "minor component" was at a higher concentration level. Not surprisingly, false alarm rates dropped when the "adulterant" became more intense. This improvement was not noticeable, however, when the major component was both familiar and when the minor component was either familiar or unpleasant— false alarms were low for the low-intensity condition and remained low at the higher level. The one case in which false alarm rates increased was the pairing of an unfamiliar major component with a familiar minor component. The reasons for this are unclear and the authors cautioned that the observations were not statistically analyzed due to small numbers of observations. In summary, the results are consistent with the notion that learning allows persons to act more analytically, and that trained experts such as perfumers are indeed able to pick out more individual notes in a complex odor blend.

V. INFORMATIONAL CONTENT OF ODORS

Reflection on the information transmission capabilities of the sense of smell leads to a recognition that we are able to differentiate a wide range of qualitatively different experiences. Many objects encountered in nature have distinct and different odors and the variation on these natural themes created by chemists for use in our everyday personal and household products would suggest no limit to the possible smells we could encounter. For example, we can experience lemon furniture polish, lemon candy, lemon dishwashing soap, and lemon soda all within a few moments of daily life. Each of these odors is a different version of lemon, and yet we are able to consistently categorize them as examples of the lemon category. How is this achieved? Several aspects of this situation have been addressed in the literature, notably the information transmission question for intensity and quality, the question of classes and categories, and the problem of consistent naming.

A. Discrimination and Channel Capacity

There appear to be very real limitations on the ability of the sense of smell to tell us much about the concentration of odorous substances in the world, once they get above threshold. Although some of this noise is surely due to concentration variation in the ambient world (or even in our laboratory stimuli), it nonetheless calls into question whether intensity information is worthy of much interest from either a functioning organism or from psychophysical investigators trying to understand smell perception. In contrast, the nose appears to be a very good chemical discriminator when it comes to differences in chemical structures and the composition of naturally occurring mixtures. It is widely demonstrated that humans can smell the difference between various enantiomers (mirror image molecules) such as the minty odor of r-carvone and the caraway odor of s-carvone. Polak et al. (1989) showed that subjects differed widely in their ability to smell the two enantiomers of ionone, some being more sensitive to one isomer and others showing the reverse pattern of sensitivity. This surprising result suggests separate receptors for the reception and discrimination of these two nearly identical molecules.

Estimates of the channel capacity of the sense of smell for different qualities have never reached a realistic asymptote. Early attempts at estimating odor quality identification found an average level of about 4 bits, or 16 odors, correctly identified from a larger set of chemical and essential oils (Engen & Pfaffmann, 1960). However, subsequent workers pointed out that the stimulus set in this early study included many simple chemicals that were possibly unfamiliar, not informationally rich, and possibly hard to discriminate. When natural objects were used for identification (lemon, pencil shavings, band-aids), performance was nearly perfect for many subjects up to the limit of the size of the set, about 80 odors (Desor & Beauchamp, 1974). This and subsequent studies suggested that the magnitude of the channel capacity for odor type was difficult to estimate (the nose of the observer might become fatigued before the actual limit was probed), and was probably very large (Cain, 1979). This fits well with common experience and the notion that we can appreciate a wide range of qualitatively different smells. The immense range of different materials available for perfumery and flavor synthesis (Bauer & Garbe, 1985; Heath, 1981) attests to the scope of qualitative differentiation by the sense of smell. Each of these materials smells different when examined "side-by-side" or in rapid succession.

B. Identification and the Olfactory–Verbal Gap

Associating an odor experience with the correct name is often difficult. In an uncued, free-association experiment, it is not uncommon for subjects to misidentify or be unable to identify about half the set of odors presented (Cain, 1979; Desor & Beauchamp, 1974; Lawless & Engen, 1977). However, when given practice with the correct names, performance will improve to near perfect levels as long as the odors are familiar and dissimilar items (Cain, 1979; Desor & Beauchamp, 1974). This has led to the speculation that experiences in the olfactory modality are unusually hard to associate with previously learned verbal labels when the task requires retrieval of that word. Evidence for this lies in the fact that multiple-choice tests of odor labeling will generally find increased levels of performance over uncued identification and will approach perfect performance in persons with a normal sense of smell when the number of choices for names is about four (Doty, 1991b). This fact is widely employed in clinical tests of smell in order to differentiate problems in olfactory perception from difficulties in name retrieval (Cain, 1979).

A second effect consistent with the notion of an olfactory–verbal gap is the "tip-of-the-nose" phenomenon (Lawless & Engen, 1977). When asked to identify a smell, subjects will often remark that they have a frustrating sense of the familiarity of the odor, but that they are unable to pin down the name of the smell. Although they are able to name similar smells or perhaps a general category for the odor, the name is elusive, a curious case of recognition without identification. This seems different from the familiar "tip-of-the-tongue" state, in which a word seems out of reach although its contextual fit in a sentence is apparent to the searcher. In the tip-of-the-nose state, a simple reading of the dictionary definition is adequate to bring to mind the correct name for the smell, whereas in the tip-of-the-tongue state, reading the dictionary definition is a common method for *eliciting* the situation, rather than curing it.

Not all authors have agreed that the problem in odor identification is simply one of accessibility or retrieval of the correct odor name. Schab and Cain (1991) argued that errors of perception and possible discriminative difficulties have not been adequately ruled out. They point to the finding by Eskenazi et al. (1986) that subjects in the triangle testing situation (picking out the odd sample from a set of three that includes one duplicate pair) often make errors, even with categorically different odors. Although average performance is quite high (87%), it still seems surprising that subjects could fail to differentiate carvone from benzaldehyde. This finding suggests that odor quality discrimination is not as efficient as is widely believed. However, it should be pointed out that the triangle procedure is more difficult than meets the eye, especially when multidimensional stimuli are used (Ennis, 1990; Ennis & Mullen, 1986; Frijters, 1979).

In contrast to this view, odor identification can sometimes be surprisingly good. Laing showed that subjects could identify singular, dissimilar odors from a known set with a single sniff—in fact, the shortest sniff they are physically able to produce (Laing, 1986). It would also seem that narrowing the field of possible odors (which without further restriction is roughly equivalent to the number of different odiferous objects!) would seem to help identification performance. Subjects in the Laing study benefitted from practice, knowledge of the set, use of common associations, and the presence of a multiple-choice matching list, all of which should facilitate circumventing the olfactory–verbal gap.

Multidimensional scaling (MDS) models from odor sorting seem quite impervious to the training levels of the panelists (Lawless & Glatter, 1990). This may reflect the insensitivity of the technique to tap into perceptual changes with experience or the fact that the method taps into simple physiologically determined or culturally shared odor quality structure. This is paralleled by findings that wine experts and nonexpert consumers do not differ much in their capacity to discriminate among wines (Solomon, 1988). However, other results show effects of experience and familiarization on simple discrimination tasks such as the detection of a low-intensity unpleasant component in a mixture (Rabin & Cain, 1989). Rabin

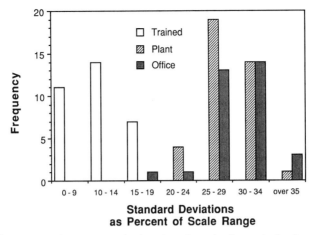

**Standard Deviations
as Percent of Scale Range**

FIGURE 2   Standard deviations on odor intensity rating scales from trained and untrained (office location, plant location) panels. Original data were from consumer product fragrances rated on the scales shown in Figure 3. Scales that had a mean value of zero from the trained panel, and therefore zero standard deviation, were omitted from all groups in this comparison. Reprinted from Lawless, H. T., Odour description and odour classification revisited. *In* Food Acceptability. Copyright 1988, pp. 27– 40, with kind permission from Elsevier Science Ltd., The Boulevard, Langford Lane, Kidlington OX5 1GB, UK.

(1988) showed that giving experience in odor labeling or semantic profiling facilitated discrimination performance in a two-interval same/different task.

Aspects of odor perception that do show interactions with experience or training usually involve recognition or semantic tasks such as odor identification or description. Rabin and Cain (1989) showed that odor recognition proceeds more effectively if the odor is familiar to the individual. Correlated with ability to recognize and with familiarity were the ability to identify the odor and to do so consistently. This gives strong evidence that performance of semantically related tasks improves dramatically with training. Perhaps this is not surprising in light of the olfactory– verbal gap. Along these lines, sensory judges trained to use general odor terms for profiling show much lower variability in ratings than do untrained consumers (Lawless, 1988) (Figure 2). They also show greater specificity and interobserver agreement in the application of terms. As shown in Figure 3, the trained panelists use only those terms that are applicable to the complex odor in question, whereas untrained observers are more variable, spreading their judgments to all available categories of description.

C. Odor Recognition Memory and Imagery

Many have noted that experience with a long-remembered but rarely experienced odor can elicit recollections with emotional overtones of places and persons from

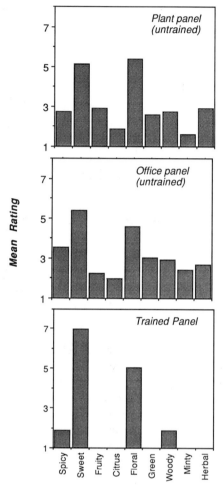

FIGURE 3   Ratings of trained and untrained fragrance evaluation panels using the descriptive scales shown on the abscissa. The trained panel gives the rating for "not present" (scale value = 1) for some odor notes not present in the product, whereas the untrained panels produce nonzero mean ratings for all scales. Reprinted from Lawless, H. T., Odour description and odour classification revisited. *In* Food Acceptability. Copyright 1988, pp. 27–40, with kind permission from Elsevier Science Ltd., The Boulevard, Langford Lane, Kidlington OX5 1GB, UK.

the past (Laird, 1935). Experimental demonstrations show that such evoked memories are indeed emotional, clear, specific, rarely thought of, and comparatively old (Herz & Cupchik, 1992). This effect led to theories that memory for odors was somehow different from memory for items encoded through other modalities. Because odors cannot be reconstructed from memory in a manner similar to verbal stimuli, the tasks for studying odor memory turned to recognition, that is, identification of an odor as one previously presented in an experimental context (as differentiated from some odors not presented).

Studies of memory over fairly short time intervals (3 to 30 sec) showed surprisingly less than perfect recognition, although the time function for decay in performance was also very flat (Engen, Kuisma, & Eimas, 1973). Long-term studies of odor recognition have also shown difficulty of encoding but resistance to forgetting (Engen & Ross, 1974; Lawless & Cain, 1975). The results of Engen et al. (1973) seemed surprising, in that subjects were asked to perform mental arithmetic tasks during the forgetting interval, a manipulation that was known to interfere with short-term storage of verbal stimuli. However, a fair question arises as to whether the verbal interference task would necessarily prohibit some type of rehearsal of the odor experience. Attempts to have subjects rehearse the to-be-recognized stimuli shed some light on the possible nature of odor rehearsal. Walk and Johns (1984) asked subjects to engage in rehearsal of the target odor name, a nontarget odor name, to smell and associate to a nontarget odor, or do whatever they pleased. Processing of the nontarget odor impaired performance, consistent with modality-specific interference that would occur if odor imagery was necessary. However, the rehearsal of a target odor name improved performance, which is also evidence for some semantic mediation (Schab & Cain, 1991). Other tasks involving semantic mediation have also enhanced recognition memory (Lyman & McDaniel, 1986). Associational learning studies of odors have also shown a tendency for subjects to produce verbally mediated schemes to enhance their performance (Lawless & Engen, 1977).

This apparently strong tendency for verbal mediation raises the still more interesting question of whether any modality-specific odor imagery exists. Cain (1977b) points out that in other sensory domains, we can operate on images. For example, in the visual domain, we are able to rotate an imagined object and in the auditory realm, we can hear a piano play any song we want. However, it is unclear whether any such active manipulation can occur with smells, although Cain speculated that some type of cognitive operations might be present in animals with highly developed senses of smell and seemingly richer olfactory lives. In humans, less than half of people asked about odor imagery report that they experience it (Schab & Cain, 1991). In general, it is unclear what people understand odor imagery to be. When the analogy is drawn to "seeing something in the mind's eye" as an example of visual imagery, the results shown in Figure 4 are obtained. Subjects report more frequent imagery for visual and auditory mental representations (H. T. Lawless, unpublished). Among those reporting imagery in each modality, the vividness of the images is also higher for auditory and visual representations. A vivid image was described as one that seemed real, complete, detailed, and had impact. Carrasco and Ridout (1993) showed that multidimensional scaling solutions of similarities of imagined and smelled odors were highly similar. This is consistent with the existence of olfactory imagery that is like olfactory perception, although some semantic mediation is still possible in their task.

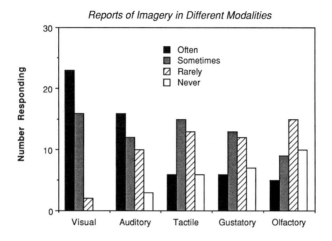

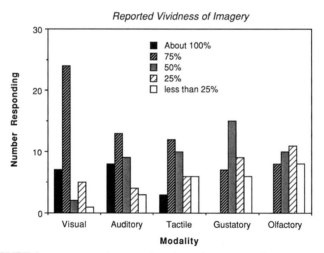

FIGURE 4   Frequency of report and vividness of imagery in different modalities.

VI. ODOR CATEGORIES

A. Impediments to Classification

Classification of odors has long been a dilemma that defies simple schemes. The most comprehensive attempt at odor classification was the treatise *Odour Description and Odour Classification* (Harper, Bate Smith, & Land, 1968). This work provided a historical review. Some approaches from multivariate behavioral methods and some partially convergent information from specialized fields such as botany were included. Those authors lamented the lack of a universal system for odor classes, analogous to primary colors or basic tastes. The most instructive aspects of the book are perhaps the impediments to a universal system that are imposed by the nature of odor quality perception.

One impediment is certainly the tendency to name odors by pointing to objects in the real world. This provides a basis for odor classification on knowledge of origin, a cognitive aspect, rather than perceptual similarity. Because the behaviorally useful information content of odors resides in how they signal the proximity of certain objects, beings, places, etc., it is understandable that odor typologies should naturally proceed from a consideration of sources. It also implies that different cultures might have different odor classes, depending on the sources for various odiferous chemicals in their locale. This also raises the problem of idiosyncratic experience—individuals as well as cultures with different experiences of an odor may produce different classification (is rose a food?).

It has been suggested that classification of odors and tastes is akin to other forms of concept learning (O'Mahony & Ishii, 1986; Lawless, 1988). The classic model for such learning is a process of abstraction, whereby an organism learns to attend to the common features that signal class membership, ignore those features or dimensions that are irrelevant, and finally generalize those sets of features and boundaries to new items so that classification can proceed with some accuracy, even for items never before encountered. Other approaches to categorization have shunned simple feature modeling. One approach points out that learning proceeds to the point at which a category prototype is abstracted (even if never seen) and that category membership is a function of similarity to that prototype (Lakoff, 1986).

Another approach stresses that no category member will have all of the attributes that define membership, but will still bear some overall "family resemblance" to other items in the set (analogous to a group of cousins in a family photo) (Rosch & Mervis, 1975). As noted in the introduction, these theories are applicable only to the extent that there is a set of abstractable features for odors. Although some highly trained odor evaluators may act analytically to decompose odor impressions, it is probably more common for people to relate to odors as unitary perceptual experiences. Thus the utility of any feature- or dimension-based model is unclear for odor perception.

Another impediment may be the reluctance of olfactory theorists to embrace a system with a large number of primaries. Murphy (1987) suggested that the tendency to prefer classification systems with seven or so primary qualities is derivative from our information-processing limitations and our discomfort with more than seven (plus or minus two) categories. Unfortunately, any system that depends on only seven superordinate classes will tend to be neither accurate nor useful (see later). Chastrette and co-workers, using cluster analysis of perfumery terms, concluded that the hierarchical structure of odor categories is probably very shallow and broad, meaning that the number of differentiable odor types is quite large (Chastrette, Elmouaffek, & Sauvegrain, 1988).

B. Hierarchies

One approach to this problem is to consider such notions of hierarchical structure as potentially simplifying. Most human knowledge systems regarding objects and perceptual qualities are hierarchical. As we learn, we come to understand the color red, then descriptions for shades of red (brick red, fire engine red), just as we learn about dogs, then types of dogs. The hierarchies are also elaborated upward and articulated as we learn rules and distinctions for superordinate groupings (mammals, vertebrates, animals). One example of a hierarchical system for odor classification is the wine aroma wheel shown in Figure 5 (Noble et al., 1987). The outer-tier terms reflect fairly specific odor experiences and the inner-tier terms apply to more general groups. The inner tiers may also be useful for odor description, because experiences of wine aroma are often somewhat vague or blended—one may know that a particular fruity smell is present, but be unable to associate it with a specific fruit. A need for the general terms is also characteristic of sensory judges in the early stages of their training. This approach to odor description can be contrasted with odor-profiling single-tier systems such as the ASTM list of 146 odor descriptors (Dravnieks, 1985). This system makes an attempt to arrange a subordinate/superordinate relationship, although some terms are clearly more general than others (spicy vs. cinnamon, meaty vs. fried chicken).

A potentially important theory concerning levels of the hierarchy is the notion of basic levels (Rosch, Mervis, Gray, Johnson, & Boyes-Braem, 1976). Rosch and co-workers proposed that a highly utilitarian level exists at which the following characteristics hold: basic levels strike a balance between within-group homoge-

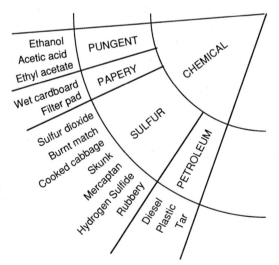

FIGURE 5   A section of the wine aroma wheel showing hierarchical arrangement of odor terms. Modified from Noble et. al. (1987).

neity and between-group differentiation; higher levels contain perceptually hetero-geneous members whereas lower level groupings become harder to differentiate from one another (e.g., types of spaniels). Basic level terms are often the first terms learned by children, the terms that come to mind most rapidly in conversation of adults. In reaction time studies, they are processed more rapidly for tasks such as verification of category membership. To date, this notion has not been applied to odor classification, although it could be argued that the outer tier of the aroma wheel is a basic level. One tentative principle for odor groups is that the superordinate categories rely on more cognitive similarity and the subordinate categories use more perceptual similarity. Moving to higher order categories will sooner or later lead to strange bedfellows.

C. Commonalities among Systems

In examining various applied systems for odor description, for example, the wine aroma wheel and the ASTM list, there are many common terms to be found. Where discrepancies occur, they can be partly explained by the specificity of one system for the particular product class or milieu such as wine aromas versus perfumery. A second explanation for discrepancies is usually on the basis of detail. Table 2 shows considerable overlap in two summaries of perfumery terms. One article was a brief note on an introductory level (Brud, 1986) and the second was an extensive cluster analysis of terms for a large array of perfume materials (Chastrette et al., 1988). Given this difference in orientation and approach, the similarity is noteworthy. A second example of coincidence is shown in Table 3. The first column shows a simple descriptive list developed using the intuition of panel leaders

TABLE 2 Terminology Comparison

Descriptive scheme	Factor analysis
Woody/nutty	Woody, nutty
Minty	Minty
Floral	Floral
Fruity (not citrus)	Fruity, noncitrus
Citrus	Citrus
Spice	Spice
Green	Green
Herbal	Caraway, anise
Sweet (vanillin, maltol)	Brown (vanilla, chocolate, molasses)
Other	Coconut, almond
	Animal, foul
	Solvent
	Burnt
	Sulfidic
	Rubber

TABLE 3   Perfumery Terms

Chastrette et al. (1988)	Brud (1986)
Miel (honey)	—
Musc (musk)	Musk
Animal	Animal
Menthe (minty)	—
Camphre (camphor)	—
Bois (woody)	Woody
Sylvestre (forest)	—
Moisi (mouldy, musty)	Fungoid
Terreux (earthy)	Earthy
Ambre	Amber
Ethere (ethereal)	Chemical
Vineux (vinous)	—
Cireux (waxy)	Aldehydic
Hesperide (citrus, orange)	—
Gras (fatty)	Fatty
Buileux (oily)	—
Epice (spicy)	Spicy
Herbace (herbal)	Herbal
Balsamique	Balsamic
Floral	Flora
Rose	—
Vert (green)	Green
Fruite	Fruity
Anise	—

to describe fragrances from consumer and household products (Civille & Lawless, 1986). The second list is derived from a factor analysis of the ASTM 146-item list as applied to several hundred flavor and fragrance materials (Jeltema & Southwick, 1986). Although there are some items not encountered (perhaps not expected) in the consumer products list that appear in the factor analysis, there is considerable similarity among common items. Taken together, these coincidences suggest that the lamentation about there being no universal system for odor classification is misplaced, if not overstated. There is quite a bit of consensus about odor types when utilitarian systems are developed for product applications. However, an exhaustive system will undoubtedly be quite complex.

D. Local Spaces

One approach to the complexity problem is to choose to study small areas of the odor space in a given experiment. The global view leads to overly simple maps of odor space with only hedonic (pleasantness) and intensity dimensions. The examination of more delimited odor types may yield more information about odor categorization and the way that odor classes shade into one another. To do this, one needs to choose a domain in which there are a suitable number of closely related odor types (Lawless, 1989; MacRae, Howgate, & Geelhoed, 1990), or construct mixtures to produce the intermediate levels or border conditions (MacRae, Rawcliffe, Howgate, & Geelhoed, 1992). Multidimensional scaling has proved to be a useful technique for examining patterns of similarity among odors (Schiffman, 1974; Schiffman, Musante, & Conger, 1978). Other procedures such as cluster analysis have also been performed (Chastrette et al., 1988; Chastrette, de Sain Laumer, & Sauvegrain, 1991) and deserve further consideration, because the best model for odor categorization may not necessarily be a multidimensional space. Tversky and Hutchinson (1986) derived criteria for the applicability of tree structures over simple euclidean spaces. The data sets they examined from the chemical senses qualified as cases in which tree structures might be more applicable than spatial solutions, especially those that have involved both some degree of nesting and some central superordinate nodes. This fits with the conclusion of Chastrette et al. that an examination of the broad range of possible odor types yields a hierarchical structure that is broad and flat, that is, has minimal nesting and a large number of highly dissimilar superordinate classes. In one comparison of nonlinear mapping with cluster analysis and nearest neighbor (tree) analysis, all methods were found to show similar structure that could be represented in a two-dimensional map (Chastrette et al., 1991). However, that analysis examined mostly superordinate general perfumery terms, for example, pyrogenous, camphoraceous, so little hierarchical structure was evident.

Nonetheless, in the case of a limited number of related odors, MDS can provide confirmatory or exploratory information on relationships and possible categories. One such study was performed to evaluate the similarities of terpene aroma materials spanning citrus and pinelike odor types, and including several materials that seemed to be intermediate to those categories (Lawless, 1989). One problem that is often encountered with odor stimuli is that the nose appears to fatigue rapidly to odors of different types presented sequentially. The usual MDS data collection procedure involves paired estimates of similarity. This is difficult to carry out for any set of odors of meaningful or interesting size. As an alternative to the pairwise similarity estimates, a sorting task was used and numbers of co-occurrences in the same sorted groupings, summed across subjects, were used as an index of odor similarity and input into MDS (Lawless, 1993; Rosenberg, Nelson, & Vivekanathan, 1968; Rosenberg & Kim, 1975).

Six odors with a priori citrus-type character, six odors with woody notes, and six odors judged to have both types of smell or a blended "ambiguous" mix of citrus and woody character were sorted. Figure 6A shows the MDS solution for these 18 odors. The validity of the solution is demonstrated by several factors: first, intuitive pairings always plotted in proximity (e.g., citral with lemon oil and cedrene with cedarwood oil). Second, mean attribute ratings for citrus and woody character

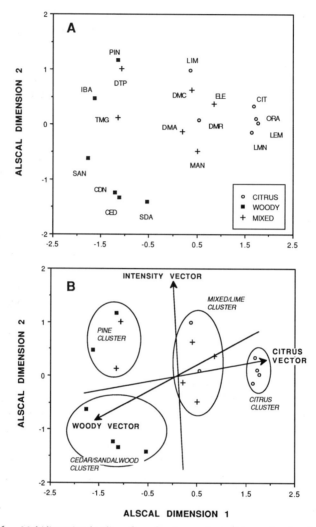

FIGURE 6   Multidimensional scaling of woody, citrus, and mixed-type odors from Lawless (1989). (A) MDS output; (B) the same plot with cluster analysis and regression of attribute vectors from rating scales. Key: CIT, citral; ORA, orange oil; LEM, lemon oil; LMN, limonene; LIM, lime oil; DMR, dimyrcetol; ELE, elemi oil; MAN, mandarin orange oil; DMC, dihydromyrcenol; DMA, dihydromyrcenyl acetate; PIN, pine oil; IBA, isobornyl acetate; TMG, tetrahydromugyl acetate; DPT, dihydroterpineol; SAN, sandalwood oil; SDA, sandela; CED, cedarwood oil; CDN, cedrene.

were regressed onto the space (Schiffman, Reynolds, & Young, 1981) and plotted in a sensible fashion as shown in Figure 6B. Third, pine odor types and the other woods plotted in separate clusters. Fourth, four of the six "ambiguous" smells plotted in an intermediate cluster. The visually apparent clusters are substantiated by cluster analysis. The solution is also hypothesis generating. The two lime-type odors, lime oil and dimyrcetol, plotted in the ambiguous group, suggesting that it is their partially woody character that differentiates them from the other citrus (lemon, orange) odors. This speculation is substantiated by their attribute ratings.

Two further observations on this approach are noteworthy. First, the solutions from sorting correspond closely to those obtained from pairwise similarity ratings (Lawless, 1993). However, the similarity ratings usually require higher dimensionality for a reasonable fit to the data, indicating a perhaps more detailed or richer qualitative structure in those comparisons (Bertino & Lawless, 1993). Second, the sorting solutions appear largely unaffected by the level of training or experience of the subjects (Lawless & Glatter, 1990). Trained panelists, college students, and consumers all give very similar groupings. This suggests some overriding cultural similarity or perhaps the sortings are tapping into fundamental, perhaps even nonverbal, physiologically based patterns of similarity.

If sorting of odors from delimited parts of the odor world can yield sensible and intuitively satisfying perceptual maps, is it possible to construct a model of the entire space using this approach? Perhaps, but it will probably require many small experiments, sorting odors from deliminated areas and studying the ways in which they "shade" into one another—in other words, a patchwork quilt of odor space.

E. Effects of Context and Contrast

Given the categorical structure described for terpene compounds of citrus and woody type, it can be asked whether category shifts occur as a function of preexposure. Such shifts as well as general contrast effects are common in the visual and auditory modalities, but have rarely been studied in smell. For example, repeated exposure to phonemes whose formant structure is on one side of a boundary [e.g., for voice onset time (/pa/)] will shift sounds that occur near the boundary into adjacent categories (to /ba/) (Eimas & Corbit, 1973). Similar shifts occur with odors. After exposure to prototypically woody odors, an ambiguous odor, dihydromyrcenol, was rated as more citruslike in character than were its ratings following citrus odors. After citrus odors, it is rated as more woody in character (Lawless, Glatter, & Hohn, 1991). To this extent, odor perception seems to operate much like situations of qualitative contrast in other modalities.

This effect could be attributed to simple sensory adaptation, particularly if dihydromyrcenol is perceived as if it were a mixture. In that case, adaptation to one component of the "mixture" (in this case one "note" of the complex odor) would cause the other to increase in intensity, the simple release-from-suppression effect described previously. However, an experiment using the reversed-pair paradigm rules this out. In the reversed-pair situation, the contextual odor is given after the target odor, to rule out any effect of preexposure or adaptation of the contextual odor on the potentially shifted item. In the case of three ambiguous citrus/woody odors, the contrast shift was still observed, even in the reversed-pair situation (Lawless et al., 1991). This is consistent with other explanations such as a change in category boundary or criterion shift, rather than a direct adaptation effect on the odor percept.

VII. ISSUES, NEEDS, AND DIRECTIONS

A. What Is the Metric for Odor Quality and the Model for Similarity?

Part of the message of Section VI was to call into question the utility of odor models based on the spatial representations that are so conveniently provided by techniques such as factor analysis and multidimensional scaling. Perhaps odor quality relationships are better captured by fractal structures or tree diagrams. In addition to such questions about the nature and characteristics of an odor quality model, we can also question the nature of our metric for similarity. One approach to specifying odor or flavor qualities is to provide a multidimensional profile, where the intensity of specific notes is rated as if the odor was composed of a set of analyzable sensory characteristics. This is typical of the descriptive analysis methods used in applied product evaluation (Caul, 1957), and also has been used as one approach to odor profiling (Dravnieks, 1985). This might be called a *psychophysical model,* because profiling is conceived of as a collection of independent intensity relationships.

Several problems leave this model unsatisfying. First, odors are not always so analyzable and often seem to form blended wholes. Even when we try to construct mixtures from cacophonous dissimilar compounds, some odor notes seem to drop out or become obscured (Laing & Francis, 1989). Second, independence on statistical grounds can be questioned in that odor characteristics are more or less correlated (Chastrette et al., 1988). Third, the task demands are such that subjects will give such profiles, whether or not they are good or veridical reflections of the odor experience. Finally, looking at the unlabeled profile may not permit recognition of the actual product (e.g., a fudgsicle), which points out that the profiling is only a weak model of the real experience (Meilgaard et al., 1991).

At the opposite extreme from this approach is the judgment of overall similarity. This would seem to avoid problems inherent in semantic selection of descriptors and in assumptions about consensus meanings for adjectives that may in fact have quite variable meanings to different people. However, the basis for similarity judgments is by no means always clear, always interpretable, always stable, or necessarily informative. Early attempts at providing multidimensional scaling solutions for highly diverse sets of odors would yield dimensions of the odor space interpretable as pleasantness and often intensity. Because these two characteristics are criteria by which any odors can be compared, they are not terribly informative regarding odor type.

Intermediate to these extremes are mixtures of the psychophysical approach and similarity scaling, wherein the task calls for an estimate of qualitative difference from a notion of an odor type as semantically labeled. One example is in the *Atlas of Odor Character Profiles* (Dravnieks, 1985), which has intensity scores for 146 descriptors, and also "applicability" scores that try to assess the degree to which each term applies to the odor character of the compound. In this extensive data set, there is a strong parallel between the percent applicability and the mean intensity ratings. The author also encountered this similarity during the training of a fragrance profile panel. After being trained to estimate the intensity of different general odor notes (spicy, fruity, etc.) in complex consumer product fragrances, the panel was instructed to rate applicability, instead of intensity. Mean profiles by either instruction were highly related. This was documented in an experiment with terpene odors of citrus, woody, and mixed (ambiguous) type as well as two mixtures of prototypical citrus and woody odorants (Lawless, 1992). In one condition, subjects were asked to rate the intensity of citrus and woody smells in each odor. A second group was instructed to evaluate the degree to which the odor was "a good example" of a citrus or a woody smell. Goodness-of-example is a common task in the study of categorical perception, especially when prototype models are being tested.

The profiles across odorants for the two methods are shown in Figure 7. The mean values are plotted in polar coordinates and connected to form polygons, to give a simple visual impression of the similarity in a geometric shape. The correlation for intensity and goodness for citrus and woody was +0.87 and +0.90, respectively. This could have several interpretations. Obviously, one of the two tasks may be an accurate reflection of how the odors are perceived, and the other task simply

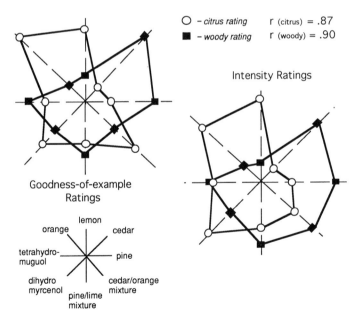

FIGURE 7 Citrus and woody intensity ratings versus citrus and woody goodness-of-example ratings for four oils, two mixed-type compounds, and two mixtures shown in the lower left. Ratings are proportional to the distance from the center point, and have been connected to form polygons. Modified from Lawless (1992).

the result of a strategy to provide a reasonable response by translating into the easier or more accurate task. Alternatively, both ratings may be derived from some more fundamental perceptual process. Third, they could both be accurate and valid, in the sense that more intense characteristics move the odor closer to the prototype of that category, and thus ratings for goodness go up along with the intensity of odor notes. Whatever the ultimate explanation, the correlation shows that there is more than one way in which to conceptualize the tasks involved in specifying odor quality.

B. Functional Significance of Specific Anosmia

A mismatch between everyday observation and the psychophysical literature occurs in the following manner: Research on smell shows strong individual differences both between individuals in the findings related to specific anosmia and in the intraindividual variation documented for changes in an individual's sensitivity over time. Yet people are able to discuss their olfactory experiences with one another much as if they are living in the same odor world. Although it is not uncommon to find that members of a flavor evaluation panel will each have some consistent strengths and weaknesses in the evaluation of specific flavor notes, the similarities among individuals following training are more striking than their differences.

This raises the question of whether the differences seen in effects such as specific anosmia have any influence in everyday life. More concretely, the effect of specific anosmia on smell sensations above thresholds and in mixtures is nearly undocumented. Several lines of thought would suggest that these effects should, in fact be potentially important. Many of the compounds to which there are specific anosmias are food flavor components (e.g., androstenone in boar taint in pork, cineole in many herbs and spices, carvone in mint, and isovaleric acid in cheeses and other fermented products). Isobutyraldehyde occurs in dairy products as a function of *Lactococcus lactis* v. *maltigenes,* imparting a malty flavor defect to some milk (Bodyfelt et al., 1988), a fact noted by Amoore in the original article on isobutyraldehyde anosmia (Amoore et al., 1976a).

As shown for individual differences in bitterness perception for phenylthiocarbamide (PTC), there is a good correlation of threshold sensitivity with above-threshold perception, particularly in the concentration ranges to which one group is highly sensitive but that are still below the threshold of most people in the insensitive group (Lawless, 1980). Whether such a correlation exists between thresholds and suprathreshold smells is still open to question. In the case of cineole and diacetyl, a reasonable correlation of the order of -0.5 to -0.6 was found (Lawless et al., 1994, 1995). The negative sign of the correlation arises because low thresholds means high sensitivity, and high reactivity above threshold. One would expect a similar or stronger relationship in the case of androstenone, because androstenone nonsmellers seem largely unaware of the smell at any concentration and would presumably give low or zero intensity ratings to a concentration considered strong by sensitive individuals. However, the size of this correlation still lags behind what is found for PTC tasting, which is of the order of -0.8, high enough to apply a suprathreshold screening procedure for determining a status of bitter tasting (Lawless, 1980).

In addition to questions about the suprathreshold intensity of smells to different individuals, we can also ask about the smell quality. O'Connell et al. (1989) and Stevens and O'Connell (1991) found a strong relationship between specific anosmia and the quality reports that were given to androstenone and pemenone, and they used this fact to support the notion that odor molecules interact with multiple perceptual channels. Sensitive individuals reported putrid-type odors whereas anosmics reported other smells. There are relatively few such observations on the effects of other anosmia-related compounds above threshold. Nonetheless, it is reasonable to suppose that if specific anosmia does represent absence of a particular receptor, then the quality of the smell to nonsensitive individuals will reflect the ability of a molecule to stimulate other receptors at higher concentrations (perhaps a receptor with lower affinity for that molecule than for other compounds to which it is more specifically tuned). When this occurs, the smell takes on the quality associated with that labeled line or pattern.

A third reason to suspect an influence of specific anosmia on suprathreshold perception includes the effects of a smell sensation on other sensations that are present. In mixtures of PTC and sucrose, most tasters will show evidence of mixture suppression whereas insensitive persons who perceive no bitterness will show no diminution of sweet taste (Lawless, 1977). In the sense of smell, increasing the intensity of one substance causes increasing mixture inhibition of a second (Cain & Drexler, 1974). Decreasing intensity through adaptation shows release-from-inhibition effects similar to those observed in taste (Lawless, 1987). Therefore one would expect a parallel reciprocity of intensity in a mixture versus inhibition of other compounds. This suggests that smelling a compound within a mixture affects other smells such as a food flavor, and that persons who lack the ability to sense that compound will also show no inhibition of other flavors, changing the general balance or overall flavor character.

However, as noted previously, the impact on everyday functioning of the specific anosmia defects requires some degree of stability over time. If individuals are as highly variable as shown in threshold retesting, then anosmic defects for particular persons may be overestimated in single experiments, and the average experience of people may not in fact differ to the extent one would suppose. As shown in the work by Wysocki and colleagues (1989), the stability or permanence of even the androstenone anosmia may be less genetically fixed than first supposed.

C. Odor Cognition and Odor Decisions

One of the truisms of perceptual testing is that given any task, human observers will find a way to optimize their performance, often employing strategies that override the intent of the investigator to isolate a mental process of interest. Historically, the nonsense syllable or trigram of Ebbinghaus was chosen to have little or no meaning to subjects in memory studies, and yet the average college sophomore is able to find associative hooks for such stimuli with ease. Parallels exist in the attempts to use nonsense visual forms (Prytulak, 1971) and in attempts to study paired associate learning for odor with arbitrary pairings of pictures and smells. In the latter case, subjects find "scenarios" to connect the pictures and smells and thus enhance later recall (Lawless & Engen, 1977).

This serves to remind us of two principles. First, basic perceptual mechanisms of smell are exceedingly difficult to isolate, especially those that one wishes to attribute to peripheral mechanisms, such as olfactory receptors. The cognitive influence of observers' strategies will always be brought to bear when subjects wish to optimize their performance, either for explicit reward, for internal gratification at their performance on what seems like a "test," or to provide the experimenter with results that the subject intuits are desired. Thus we find Engen's lamentation that the signal detection methods should be further utilized in a perceptual system that is both noisy and prone to effects of observer bias. Second, the interplay of higher cognitive processes, particularly semantic correlations such as recognition and codability, can be seen if the experiment is careful to account for individual associations (Rabin & Cain, 1989). The extent to which higher cognitive effects may sharpen odor perception is open to further study, although the effects of experience on tasks such as odor profiling or picking out a contaminant from a mixture suggest that cognitive training is an important part of olfactory learning. Verbal processes and olfactory discriminative abilities go hand in hand. In cases of accidents such as gas explosions when the odor was apparently undetected, one can usually find a number of situational factors that influenced a decision resulting in the odor being unattended to or discounted. This reminds us that olfactory-guided behaviors are not only determined by how well the nose is working, but also by how well the brain that is hooked to the nose is working.

References

Amoore, J. E. (1971). Olfactory genetics and anosmia. In L. M. Beidler (Ed.), *Handbook of sensory physiology* (pp. 245–256). Berlin: Springer-Verlag.

Amoore, J. E. (1977). Specific anosmia and the concept of primary odors. *Chemical Senses and Flavor, 2,* 267–281.

Amoore, J. E., & Forrester, L. J. (1976). The specific anosmia to trimethylamine: The fishy primary odor. *Journal of Chemical Ecology, 2,* 49–56.

Amoore, J. E., Forrester, L. J., & Buttery, R. G. (1975). Specific anosmia to 1-pyrroline: The spermous primary odor. *Journal of Chemical Ecology, 1,* 299–310.

Amoore, J. E., Forrester, L. J., & Pelosi, P. (1976a). Specific anosmia to isobutyraldehyde: The malty primary odor. *Chemical Senses, 2,* 17–25.

Amoore, J. E., Pelosi, P., & Forrester, L. J. (1976b). Specific anosmias to 5α-androst-16-en-3-one and ω-pentadecalactone: The urinous and musky primary odors. *Chemical Senses and Flavor, 2,* 401–425.

Amoore, J. E., & Steinle, S. (1991). A graphic history of specific anosmia. In C. J. Wysocki & M. R. Kare (Eds.), *Chemical senses. Genetics of perception and communication* (pp. 331–351). New York: Dekker.

Amoore, J. E., Venstrom, D., & Davis, A. R. (1968). Measurement of specific anosmia. *Perceptual and Motor Skills, 26,* 143–164.

Anderson, N. H. (1970). Functional measurement and psychophysical judgement. *Psychological Review,* 77, 153–170.

Anderson, N. H. (1977). Note on functional measurement and data analysis. *Perception & Psychophysics, 21,* 201–215.

Antinone, M. A., Lawless, H. T., Ledford, R. A., & Johnston, M. (1994). The importance of diacetyl as a flavor component in full fat cottage cheese. *Journal of Food Science, 59,* 38–42.

ASTM (1978). *Compilation of odor and taste threshold values data.* Philadelphia: American Society for Testing and Materials.

ASTM (1991). Standard practice for determination of odor and taste threshold by a forced-choice method of limits. In *Annual book of standards* (pp. 34–39). Philadelphia: American Society for Testing and Materials.

Bartoshuk, L. M. (1975). Taste mixtures: Is mixture suppression related to compression? *Psychology & Behavior, 14,* 643–649.

Bartoshuk, L. M., & Cleveland, C. T. (1977). Mixtures of substances with similar tastes: A test of a new model of taste mixture interactions. *Sensory Processes, 1,* 177–186.

Bauer, K., & Garbe, D. (1985). *Common fragrance and flavor materials.* Weinheim, Germany: VCH Verlag.

Baydar, A., Petrzilka, M., & Schott, M.-P. (1993). Olfactory thresholds for androstenone and galaxolide: Sensitivity, insensitivity and specific anosmia. *Chemical Senses, 18,* 661–668.

Beets, M. G. J. (1983). Odor and stimulant structure. In E. T. Theimer (Ed.), *Fragrance chemistry* (pp. 77–122). New York: Academic Press.

Berglund, B., Berglund, U., Ekman, G., & Engen, T. (1971a). Individual psychophysical functions for 28 odorants. *Perception & Psychophysics, 9,* 379–384.

Berglund, B., Berglund, U., Engen, T., & Lindvall, T. (1971b). The effect of adaptation and odor detection. *Perception & Psychophysics, 9,* 435–438.

Berglund, B., Berglund, U., Lindvall, T., & Svensson, L. T. (1973). A quantitative principle of perceived odor intensity in odor mixtures. *Journal of Experimental Psychology, 100,* 29–38.

Bertino, M., Lawless, H. T. (1993). Understanding mouthfeel attributes: A multidimensional scaling approach. *Journal of Sensory Studies, 8,* 101–114.

Bodyfelt, F. W., Tobias, J., & Trout, G. M. (1988). *Sensory evaluation of dairy products.* New York: Van Nostrand/AVI.

Brennand, C. P., Ha, J. K., & Lindsay, R. C. (1989). Aroma properties and thresholds of some branched chain and other minor volatile fatty acids occurring in milkfat and meat lipids. *Journal of Sensory Studies, 4,* 105–120.

Brown, K. S., Maclean, C. M., & Robinette, R. R. (1968). The distribution of the sensitivity of chemical odors in man. *Human Biology, 40,* 456–472.

Brud, W. S. (1986). Words versus odors: How perfumers communicate. *Perfumer & Flavorist, 11,* 27–44.

Burgard, D. R., & Kuznicki, J. T. (1990). *Chemometrics: chemical and sensory data.* Boca Raton, FL: CRC Press.

Buttery, R. G., Teranishi, R., Flath, R. A., & Ling, L. C. (1989). Fresh tomato volatiles. In R. G. Buttery & T. Teranishi (Eds.), *Flavor chemistry* (pp. 213). Washington, DC: American Chemical Society.

Cain, W. S. (1974). Perception of odor intensity and the time-course of olfactory adaptation. *ASHRAE Transactions, 80,* 53–75.

Cain, W. S. (1975). Odor intensity: Mixtures and masking. *Chemical Senses and Flavour, 1,* 339–352.

Cain, W. S. (1976). Olfaction and the common chemical sense: Some psychophysical contrasts. *Sensory Processes, 1,* 57–67.

Cain, W. S. (1977a). Differential sensitivity for smell: "Noise" at the nose. *Science, 195*(25 February), 796–798.

Cain, W. S. (1977b). Physical and cognitive limitations on olfactory processing in human beings. In D. Müller-Schwarze & M. M. Mozell (Eds.), *Chemical communication in vertebrates* (pp. 287–302). New York: Plenum.

Cain, W. S. (1979). To know with the nose: Keys to odor identification. *Science, 203,* 467–470.

Cain, W. S., & Drexler, M. (1974). Scope and evaluation of odor counteraction and masking. *Annals of the New York Academy of Sciences, 237,* 427–439.

Cain, W. S., & Engen, T. (1969). Olfactory adaptation and the scaling of odor intensity. In C. Pfaffmann (Ed.), *Olfaction and taste III* (pp. 127–141). New York: Rockefeller Univ. Press.

Cain, W. S., & Murphy, C. L. (1980). Interaction between chemoreceptive modalities of odor and irritation. *Nature, 284,* 255–257.

Carrasco, M., & Ridout, J. B. (1993). Olfactory perception and olfactory imagery: A multidimensional analysis. *Journal of Experimental Psychology: Human Perception and Performance, 19,* 287–301.

Caul, J. F. (1957). The profile method of flavor analysis: *Advances in Food Research, 7,* 1–40.

Chastrette, M., de Saint Laumer, J.-Y., & Sauvegrain, P. (1991). Analysis of a system of description of odors by means of four different multivariate statistical methods. *Chemical Senses, 16*(1), 81–93.

Chastrette, M., Elmouaffek, E., & Sauvegrain, P. (1988). A multidimensional statistical study of similarities between 74 notes used in perfumery. *Chemical Senses, 13,* 295–305.

Chastrette, M., & Zakarya, D. (1991). Molecular structure and smell. In D. G. Laing, R. L. Doty, & W. Breipohl (Eds.), *The human sense of smell* (pp. 77–92). Berlin: Springer-Verlag.

Civille, G. L., & Lawless, H. T. (1986). The importance of language in describing perceptions. *Journal of Sensory Studies, 1,* 203–215.

Cliff, M., & Heymann, H. (1992). Descriptive analysis of oral pungency. *Journal of Sensory Studies, 7,* 279–290.

Cometto-Muñiz, J. E., & Hernandez, S. M. (1990). Odorous and pungent attributes of mixed and unmixed odorants. *Perception & Psychophysics, 47,* 391–399.

Desor, J. A., & Beauchamp, G. K. (1974). The human capacity to transmit olfactory information. *Perception & Psychophysics, 16,* 551–556.

Dorries, K. M., Schmidt, H. J., Beauchamp, G. K., & Wysocki, C. J. (1989). Changes in the sensitivity to the odor of androstenone during adolescence. *Developmental Psychobiology, 22,* 423–435.

Doty, R. L. (1991a). Psychophysical measurement of odor perception in humans. In D. G. Laing, R. L. Doty, & W. Breipohl (Eds.), *The human sense of smell* (pp. 95–143). Berlin: Springer-Verlag.

Doty, R. L. (1991b). Olfactory system. In T. V. Getchell, L. M. Bartoshuk, R. L. Doty, & J. B. Snow (Eds.), *Smell and taste in health and disease* (pp. 175–203). New York: Raven.

Dravnieks, A. (1985). *Atlas of odor character profiles.* Philadelphia: American Society for Testing and Materials.

Eimas, P. D., & Corbit, T. E. (1973). Selective adaptation of linguistic feature detectors. *Cognitive Psychology, 4,* 99–109.

Ekman, G., Berglund, B., Berglund, U., & Lindvall, T. (1967). Perceived intensity of odor as a function of time. *Scandinavian Journal of Psychology, 8,* 177–186.

Engen, T. (1960). Effects of practice and instruction on olfactory thresholds. *Perceptual and Motor Skills, 10,* 195–198.

Engen, T. (1982). *The perception of odors.* New York: Academic Press.

Engen, T., Kuisma, J. E., & Eimas, P. D. (1973). Short-term memory for odors. *Journal of Experimental Psychology, 99,* 222–225.

Engen, T., & Lindström, C. (1962). *The effects of adaptation on odor mixtures.* (Reports from the Psychological Laboratory Number 128). Univ. of Stockholm.

Engen, T., & Pfaffmann, C. (1959). Absolute judgments of odor intensity. *Journal of Experimental Psychology, 58,* 23–26.

Engen, T., & Pfaffmann, C. (1960). Absolute judgments of odor quality. *Journal of Experimental Psychology, 59,* 214–219.

Engen, T., & Ross, B. R. (1974). Long term memory for odors. *Journal of Experimental Psychology, 100,* 221–227.

Ennis, D. M. (1990). Relative power of difference testing methods in sensory evaluation. *Food Technology, 44*(4), 114, and 116–117.

Ennis, D. M. (1993). The power of sensory discrimination methods. *Journal of Sensory Studies, 8,* 353–370.

Ennis, D. M., & Mullen, K. (1986). Theoretical aspects of sensory discrimination. *Chemical Senses, 11,* 513–522.

Erickson, R. P., & Covey, E. (1980). On the singularity of taste sensations: What is a taste primary? *Psychology & Behavior, 25,* 527–533.

Eskenazi, B., Cain, W. S., & Friend, K. (1986). Exploration of olfactory aptitude. *Bulletin of the Psychonomic Society, 24,* 203–206.

Frank, R. A., Mize, S. J., & Carter, R. (1989). An assessment of binary mixture interactions for nine sweeteners. *Chemical Senses, 14,* 621–632.

Frijters, J. E. R. (1978). A critical analysis of the odour unit number and its use. *Chemical Senses & Flavour, 3,* 227–233.

Frijters, J. E. R. (1979). The paradox of the discriminatory nondiscriminators resolved. *Chemical Senses, 4,* 355–358.

Geldard, F. A. (1972). *The human senses* (2nd ed.). New York: Wiley.

Gibson, J. J., & Gibson, E. J. (1955). Perceptual learning: Differentiation or enrichment. *Psychological Review, 62,* 32–41.

Green, B. G., & Lawless, H. T. (1991). The psychophysics of somatosensory chemoreception in the nose and mouth. In L. M. B. T. V. Getchell & J. B. Snow (Eds.), *Smell and Taste in Health and Disease* (pp. 235–253). New York: Raven.

Gregson, R. A. M. (1986). Qualitative and aqualitative intensity components of odor mixtures. *Chemical Senses, 11,* 455–470.

Halpern, B. P. (1991). More than meets the tongue: Temporal characteristics of taste intensity and quality. In H. T. L. A. B. P. Klein (Ed.), *Sensory science theory and applications in foods* (pp. 37–105). New York: Dekker.

Harper, R., Bate Smith, E. C., & Land, D. G. (1968). *Odour description and odour classification: A multidisciplinary examination.* New York: Elsevier.

Heath, H. B. (1981). *Source books of flavors.* Westport, CT: AVI Publishing.

Herz, R. S., & Cupchik, G. C. (1992). An experimental characterization of odor-evoked memories in humans. *Chemical Senses, 17,* 519–528.

Jeltema, M. A., & Southwick, E. W. (1986). Evaluation and applications of odor profiling. *Journal of Sensory Studies, 1,* 123–136.

Köster, E. P. (1969). Intensity in mixtures of odorous substances. In C. Pfaffmann (Ed.), *Olfaction and taste III* (pp. 142–149). New York: Rockefeller Univ. Press.

Köster, E. P., & de Wijk, R. A. (1991). Olfactory adaptation. In D. G. Laing, R. L. Doty, & W. Briepohl (Eds.), *The human sense of smell* (pp. 199–215). Berlin: Springer-Verlag.

Kroeze, J. H. A., & Bartoshuk, L. M. (1985). Bitterness suppression as revealed by split-tongue taste stimulation in humans. *Physiology and Behavior, 35,* 779–783.

Laffort, P., & Dravnieks, A. (1982). Several models of suprathreshold quantitative olfactory interaction in humans applied to binary, ternary and quaternary mixtures. *Chemical Senses, 7,* 153–174.

Laffort, P., Etcheto, M., Patte, F., & Marfaing, P. (1989). Implications of power law exponent in synergy and inhibition of olfactory mixtures. *Chemical Senses, 14,* 11–23.

Laing, D. G. (1986). Identification of single dissimilar odors is achieved by humans with a single sniff. *Physiology & Behavior, 37*(1), 163–170.

Laing, D. G. (1991). Characteristics of the human sense of smell when processing odor mixtures. In D. G. Laing, R. L. Doty, & W. Breipohl (Eds.). *The human sense of smell* (pp. 241–259). Berlin: Springer-Verlag.

Laing, D. G., & Francis, G. W. (1989). The capacity of humans to identify odors in mixtures. *Physiology & Behavior, 46,* 809–814.

Laing, D. G., Panhuber, H., Wilcox, M. E., & Pitmann, E. A. (1984). Quality and intensity of binary odor mixtures. *Physiology & Behavior, 33,* 309–319.

Laing, D. G., & Willcox, M. E. (1983). Perception of components in binary odor mixtures. *Chemical Senses, 7,* 249–264.

Laird, D. A. (1935). What can you do with your nose? *Scientific Monthly, 41,* 126–130.

Lakoff, G. (1986). *Women, fire and dangerous things: What categories tell us about the nature of thought.* Chicago: Chicago Univ. Press.

Laska, M., & Hudson, R. (1992). Ability to discriminate between related odor mixtures. *Chemical Senses, 17,* 403–415.

Lawless, H. T. (1977). The pleasantness of mixtures in taste and olfaction. *Sensory Processes, 1,* 227–237.

Lawless, H. T. (1980). A comparison of different methods for assessing sensitivity to the taste of phenylthiocarbamide (PTC). *Chemical Senses, 5,* 247–256.

Lawless, H. T. (1984). Oral chemical irritation: Psychophysical properties. *Chemical Senses, 9,* 143–155.

Lawless, H. T. (1987). An olfactory analogy to release from mixture suppression in taste. *Bulletin of the Psychonomic Society, 25,* 266–268.

Lawless, H. T. (1988). Odour description and odour classification revisited. In D. M. H. Thomson (Ed.), *Food acceptability* (pp. 27–40). London: Elsevier.

Lawless, H. T. (1989). Exploration of fragrance categories and ambiguous odors using multidimensional scaling and cluster analysis. *Chemical Senses, 14,* 349–360.

Lawless, H. T. (1991). Effects of odors on mood and behavior. In D. G. Laing, R. L. Doty, & W. Breipohl (Eds.), *The human sense of smell* (pp. 361–386). Berlin: Springer-Verlag.

Lawless, H. T. (1992). Unexpected congruence in odor quality and intensity ratings. *Chemical Senses, 17,* 657–658.

Lawless, H. T. (1993). Characterization of odor quality through sorting and multidimensional scaling. In C. H. Manley & C. T. Ho (Eds.), *Flavor measurement* (pp. 159–183). New York: Dekker.

Lawless, H. T. (1996). Flavor. In E. C. Carterette & M. P. Friedman (Eds.) *Handbook of perception and cognition, Vol. 16, cognitive ecology.* San Diego: Academic Press.

Lawless, H. T., Antinone, M. J., Ledford, R. A. & Johnston, M. (1994). Olfactory responsiveness to diacetyl. *Journal of Sensory Studies, 9,* 47–56.

Lawless, H. T., & Cain, W. S. (1975). Recognition memory for odors. *Chemical Senses and Flavor, 1,* 331–337.

Lawless, H. T., Corrigan Thomas, C. L., & Johnston, M. (1995). Variation in odor thresholds for L-carvone and cineole and correlations with suprathreshold intensity ratings. *Chemical Senses, 20,* 9–17.

Lawless, H. T., & Engen, T. (1977). Associations to odors: Interference, mnemonics and verbal labeling. *Journal of Experimental Psychology: Human Learning and Memory, 3,* 52–57.

Lawless, H. T., & Glatter, S. (1990). Consistency of multidimensional scaling models derived from odor sorting. *Journal of Sensory Studies, 2,* 217–230.

Lawless, H. T., Glatter, S., & Hohn, C. (1991). Context dependent changes in the perception of odor quality. *Chemical Senses, 16,* 349–360.

Lawless, H. T., & Malone, G. J. (1986). Comparisons of rating scales: Sensitivity, replicates and relative measurement. *Journal of Sensory Studies, 1,* 155–174.

Lehrer, A. (1983). *Wine and conversation.* Bloomington, IN: Indiana Univ. Press.

Lundahl, D. S., Lukes, B. K., McDaniel, M. R., & Henderson, L. A. (1986). A semi-ascending paired difference method for determining the threshold of added substances to background media. *Journal of Sensory Studies, 1,* 291–306.

Lyman, B. J., & McDaniel, M. A. (1986). Effects of encoding strategy on long-term memory for odors. *Quarterly Journal of Experimental Psychology, 38A,* 753–765.

MacRae, A. W., Howgate, P., & Geelhoed, E. (1990). Assessing the similarity of odours by sorting and by triadic comparison. *Chemical Senses, 15,* 691–699.

MacRae, A. W., Rawcliffe, T., Howgate, P., & Geelhoed, E. (1992). Patterns of odour similarity among carbonyls and their mixtures. *Chemical Senses, 17,* 119–125.

MacMillan, N. A., & Creelman, C. D. (1991). *Detection theory: A users guide.* Cambridge: Cambridge Univ. Press.

Marin, A. B., Barnard, J., Darlington, R. B., & Acree, T. E. (1991). Sensory thresholds: Estimation from dose-response curves. *Journal of Sensory Studies, 6*(4), 205–225.

McBride, R. L. (1987). Taste psychophysics and the Beidler equation. *Chemical Senses, 12,* 323–332.

Meilgaard, M., Civille, G. V., & Carr, B. T. (1991). *Sensory evaluation techniques* (2nd ed.). Boca Raton, FL: CRC Press.

Moio, L., Langlois, D., Etievant, P. X., & Addeo, F. (1993). Powerful odorants in water buffalo and bovine mozzarella cheese by use of extract dilution sniffing analysis. *Italian Journal of Food Science, 3,* 227–237.

Moskowitz, H. R., & Barbe, C. D. (1977). Profiling of odor components and their mixtures. *Sensory Processes, 1,* 212–226.

Moskowitz, H. R., Dravnieks, A., Cain, W. S., & Turk, A. (1974). Standardized procedure for expressing odor intensity. *Chemical Senses and Flavor, 1,* 235–237.

Mozell, M. M., Kent, P. F., Scherer, P. W., Hornung, D. E., & Murphy, S. J. (1991). Nasal airflow. In T. V. Getchell, L. M. Bartoshuk, R. L. Doty, & J. B. Snow (Eds.), *Smell and taste in health and disease* (pp. 481–492). New York: Raven.

Murphy, C. (1987). Olfactory psychophysics. In T. E. Finger & W. L. Silver (Eds.), *Neurobiology of taste and smell* (pp. 251–273). New York: Wiley.

Noble, A. C., Arnold, R. A., Buechsenstein, J., Leach, E. J., Schmidt, J. O., & Stern, P. M. (1987). Modification of a standardized system of wine aroma terminology. *American Journal of Enology and Viticulture, 38*(2), 143–146.

O'Connell, R. J., Stevens, D. A., Akers, R. P., Coppola, D. M., & Grant, A. J. (1989). Individual differences in the quantitative and qualitative responses of human subjects to various odors. *Chemical Senses, 14,* 293–302.

O'Mahony, M., & Ishii, R. (1986). Umami taste concept: Implications for the dogma of four basic tastes. In Y. Kawamura & M. R. Kare (Eds.), *Umami: A basic taste* (pp. 75–93). New York: Dekker.

O'Mahony, M., Atassi-Sheldon, S., Rothman, L., & Murphy-Ellison, T. (1983). Relative singularity/mixedness judgements for selected taste stimuli. *Physiology & Behavior, 31,* 749–755.

Overbosch, P. (1986). A theoretical model for perceived intensity in human taste and smell as a function of time. *Chemical Senses, 11,* 315–329.

Patte, F., & Laffort, P. (1979). An alternative model of olfactory quantitative interaction in binary mixtures. *Chemical Senses and Flavour, 4*(4), 267–274.

Patterson, M. Q., Stevens, J. C., Cain, W. S., & Cometto-Muñiz, J. E. (1993). Detection thresholds for an olfactory mixture and its three constituent compounds. *Chemical Senses, 18,* 723–734.

Pelosi, P., & Pisanelli, A. M. (1981). Specific anosmia to 1,8-cineole: The camphor primary odor. *Chemical Senses, 6,* 87–93.

Pelosi, P., & Viti, R. (1978). Specific anosmia to *I*-carvone: The minty primary odour. *Chemical Senses and Flavour, 3,* 331–337.

Polak, E. H., Fombon, A. M., Tilquin, C., & Punter, P. H. (1989). Sensory evidence for olfactory receptors with opposite chiral selectivity. *Behavioral Brain Research, 31,* 199–206.

Prytulak, L. S. (1971). Natural language mediation. *Cognitive Psychology, 2,* 1–56.

Punter, P. H. (1983). Measurement of human olfactory thresholds for several groups of structurally related compounds. *Chemical Senses, 7,* 215–235.

Rabin, M. D. (1988). Experience facilitates olfactory quality discrimination. *Perception & Psychophysics, 44,* 532–540.

Rabin, M. D., & Cain, W. S. (1986). Determinants of measured olfactory sensitivity. *Perception & Psychophysics, 39*(4), 281–286.

Rabin, M. D., & Cain, W. S. (1989). Attention and learning in the perception of odor mixtures. In D. G. Laing, W. S. Cain, R. L. McBride, & B. W. Ache (Eds.), *Perception of complex smells and tastes* (pp. 173–188). Sidney, Australia: Academic Press.

Rehn, T. (1978). Perceived odor intensity as a function of airflow through the nose. *Sensory Processes, 2,* 198–205.

Rosch, E., & Mervis, C. B. (1975). Family resemblance: Studies in the internal structure of categories. *Cognitive Psychology, 7,* 573–605.

Rosch, E., Mervis, C. B., Gray, W. D., Johnson, D. M., & Boyes-Braem, P. (1976). Basic objects in natural categories. *Cognitive Psychology, 8,* 382–439.

Rosenberg, S., & Kim, M. P. (1975). The method of sorting data as a gathering procedure in multivariate research. *Multivariate Behavioral Research, 10,* 489–502.

Rosenberg, S., Nelson, C., & Vivekanathan, P. S. (1968). A multidimensional approach to the structure of personality impressions. *Journal of Personality and Social Psychology, 9,* 283–294.

Rozin, P. (1982). "Taste–smell confusions" and the duality of the olfactory sense. *Perception & Psychophysics, 31,* 397–401.

Schab, F. R., & Cain, W. S. (1991). Memory for odors. In D. G. Laing, R. L. Doty, & W. Breipohl (Eds.), *The human sense of smell* (pp. 217–240). Berlin: Springer-Verlag.

Schiffman, S. S. (1974). Physicochemical correlates of olfactory quality. *Science, 185,* 112–117.

Schiffman, S. S., Musante, G., & Conger, J. (1978). Application of multidimensional scaling to ratings of foods for obese and normal weight individuals. *Physiology & Behavior, 21,* 417–422.

Schiffman, S. S., Reynolds, M. L., & Young, F. W. (1981). *Introduction to multidimensional scaling.* New York: Academic Press.

Smith, L. B., & Kemler, D. G. (1977). Developmental trends in free classification: Evidence for a new conceptualization of perceptual development. *Journal of Experimental Child Psychology, 24,* 297–298.

Solomon, G. E. A. (1988). *Great expectorations: The psychology of expert wine talk.* Doctoral Dissertation, Cambridge, MA: Harvard University.

Solomon, G. E. A. (1990). The psychology of novice and expert wine talk. *American Journal of Psychology, 103,* 495–517.

Stevens, D. A., & O'Connell, R. J. (1991). Individual differences in thresholds and quality reports of human subjects to various odors. *Chemical Senses, 16,*(1), 57–67.

Stevens, J. C., Cain, W. S., & Burke, R. J. (1988). Variability of olfactory thresholds. *Chemical Senses, 13,* 643–653.

Stevens, J. C., & Dadarwala, A. D. (1993). Variability of olfactory thresholds and its role in assessment of aging. *Perception & Psychophysics, 54,* 296–302.

Stevens, S. S. (1962). The surprising simplicity of sensory metrics. *American Psychologist, 17,* 29–39.

Stevens, S. S., & Galanter, E. H. (1957). Ratio scales and category scales for a dozen perceptual continua. *Journal of Experimental Psychology, 54,* 377–411.

Sühnel, J. (1993). Evaluation of interaction in olfactory and taste mixtures. *Chemical Senses, 18,* 131–149.

Taylor, R. (1994). Brave new nose: Sniffing out human sexual chemistry. *Journal of NIH Research, January,* 47–51.

Thompson, R. H., & Pearson, A. M. (1977). Quantitative determination of 5-androst-16-en-3-one by gas chromatography-mass spectrometry and its relationship to sex odor intensity of pork. *Journal of Agricultural and Food Chemistry, 25,* 1241–1245.

Tversky, A., & Hutchinson, J. W. (1986). Nearest neighbor analysis of psychological spaces. *Psychological Review, 93,* 3–22.

Walk, H. A., & Johns, E. E. (1984). Interference and facilitation in short term memory for odors. *Perception & Psychophysics, 36,* 508–514.

Whissell-Buechy, D., & Amoore, J. E. (1973). Odour-blindness to musk: Simple recessive inheritance. *Nature, 242,* 271–272.

Wysocki, C. J., & Beauchamp, G. K. (1988). Ability to smell androstenone is genetically determined. *Proceedings of the National Academy of Sciences, U.S.A., 81,* 4899–4902.

Wysocki, C. J., & Beauchamp, G. K. (1991). Individual differences in human olfaction. In C. J. Wysocki & M. R. Kare (Eds.), *Chemical senses. Genetics of perception and communication* (pp. 353–373). New York: Dekker.

Wysocki, C. J., Dorries, K. M., & Beauchamp, G. K. (1989). Ability to perceive androstenone can be acquired by ostensibly anosmic people. *Proceedings of the National Academy of Sciences, U.S.A., 86,* 7976–7978.

Wysocki, C. J., Pierce, J. D., & Gilbert, A. N. (1991). Geographic, cross-cultural, and individual variation in human olfaction. In T. V. Getchell, L. M. Bartoshuk, R. L. Doty, & J. B. Snow (Eds.), *Smell and taste in health and disease* (pp. 287–314). New York: Raven.

CHAPTER 5

Clinical Disorders of Smell and Taste

Beverly J. Cowart
I. M. Young
Roy S. Feldman
Louis D. Lowry

I. INTRODUCTION

Throughout most of this century, disorders of smell and taste have received relatively little attention from the medical community. This was not always the case, however. For example, in an 1884 medical text on diseases of the nose and throat, Mackenzie devoted as much space to a consideration of chemosensory (primarily smell) dysfunction as to allergic rhinitis, citing medical references going back to 1700, and well over a dozen case reports and general articles on this topic that had appeared in the 19th century medical literature to that point.

The past decade has seen a renewed medical interest in disorders of these so-called minor senses. Although chemosensory dysfunction is not as obvious to the observer as is dysfunction in vision or audition, nor does it have the broad life-style implications, it can impact substantially on quality of life, impede performance in some occupations (e.g., food preparation, perfumery), lead to nutritional difficulties (Mattes & Cowart, 1994; Mattes et al., 1990; Mattes-Kulig & Henkin, 1985), and render individuals more vulnerable to the hazards of fire, chemical toxins in the environment, and spoiled food. Moreover, it is becoming increasingly clear that these types of dysfunction are not rare, but affect a substantial portion of the population at some point in their lives. At the very least, many if not most individuals experience measurable loss in olfactory sensitivity with aging (Cain & Stevens, 1989; Cowart, 1989; Doty, Shaman, Applebaum, et al., 1984; Gilbert & Wysocki, 1987; Ship & Weiffenbach, 1993; Stevens & Dadarwala, 1993). In addition, a number of common medical conditions are associated with long-term, permanent, or recurrent chemosensory dysfunction. For example, it has been shown that clinically significant diminutions in olfactory sensitivity are present in over 23% of patients suffering from allergic rhinitis (Cowart, Flynn-Rodden, McGeady, & Lowry, 1993). Because nasal allergies are estimated to afflict 10–15% of the general population (Fadal, 1987; Seebolm, 1978), 2.5–3.5% of the population could be expected to suffer from smell loss as a result of this etiologic factor alone.

One manifestation of the renewed clinical interest in the chemical senses has been the appearance of numerous reviews of chemosensory disorders in medical journals and texts in the past 15 years (e.g., Estrem & Renner, 1987; Mott & Leopold, 1991; Schiffman, 1983a, 1983b; Scott, 1989; Smith, 1991; Snow, 1983; and see Getchell, Doty, Bartoshuk, & Snow, 1991). It would be impossible not to replicate those efforts to some extent in a chapter on this topic. In an attempt to present the subject in a somewhat different perspective, however, the present review will focus more on unanswered questions, and tentative but intriguing observations made in clinical settings, than on providing a comprehensive overview of the conditions and medications that have been associated with chemosensory dysfunction. In most cases, the observations will be drawn from the authors' own experiences in the Monell–Jefferson Chemosensory Clinical Research Center (MJC) and those of researchers affiliated with the University of Pennsylvania Smell and Taste Center (UPenn), and the Connecticut Chemosensory Clinical Research Center (CCCRC). In addition, repeated reference will be made to the Mackenzie 1884 text in order to compare and contrast our current understanding of chemosensory dysfunction with that of a century ago.

II. SMELL VERSUS TASTE: CONFUSION AND RELATIVE VULNERABILITIES

"The recognition of the bitter, sweet, salt, and acid characters of food by the tongue and fauces constitutes taste. The appreciation of the *savor* of meat, the *flavor* of fruit, and the *bouquet* of wine, depends entirely on smell. It is necessary to call attention to these facts, because the mistake is not unfrequently [*sic*] made by medical writers, of describing cases as loss of taste, when it is clear from the context that they mean loss of smell" (Mackenzie, 1884, p. 323). Although medical writers of today are probably somewhat more sensitive to this distinction, the confusion Mackenzie describes is still commonly observed among patients presenting with chemosensory complaints, underscoring the need for careful sensory evaluation of these complaints (Deems et al., 1991; Gent, Goodspeed, Zagraniski, & Catalanotto, 1987; Goodspeed, Gent, & Catalanotto, 1987; Mott & Leopold, 1991). Mackenzie goes on to remark that "while taste is rarely impaired, smell is often altogether lost" (p. 323), a point that has been largely confirmed in the major chemosensory clinical centers.

For example, figures from UPenn (Deems et al., 1991) indicate that although approximately 66% of their patients present with a complaint of taste loss ($N =$

750), fewer than 4% are found to have a measurable gustatory deficit; in contrast, 71% are found to have absent or diminished smell function. Similarly, at MJC 65.7% of 833 patients have presented with a taste loss complaint, but we have found only 8.8% to suffer from a clinically significant taste deficit, and only 2 to have a complete loss, compared with almost 67% found to have measurable smell dysfunction (31.8% are considered to suffer from a complete loss). The frequency with which taste loss is diagnosed at the CCCRC (~30% of 441 cases, with 2 cases of ageusia) is substantially higher than at either UPenn or MJC, but still much lower than the frequency of either complaints of taste loss (~60%) or diagnosed smell deficits (which is also somewhat higher than at the other two centers: 86%, with 51% considered to suffer from anosmia) (Goodspeed et al., 1987).

A cursory consideration of the anatomies of these two sensory systems provides one obvious explanation for the apparent difference in their relative vulnerabilities. Whereas olfaction is subserved by a single cranial nerve (I), branches of three cranial nerves (VII, IX, and X) carry gustatory information. Moreover, the olfactory nerve is located in a somewhat vulnerable position in that the axons must pass through the cribriform plate of the ethmoid bone prior to dissemination on the surface of the olfactory bulb. They are therefore potentially subject to tearing or severing as a result of coup contra coup forces that may be associated with head injury (Costanzo & Zasler, 1991).

In addition, olfactory receptors are localized in a relatively small patch of tissue high in the nasal cavity, and any number of factors producing changes in nasal patency or air flow patterns could potentially limit the access of stimulus molecules to those receptors. In contrast, gustatory receptors are found on a large portion of the tongue dorsum, and a significant number may also be found on the soft palate, larynx, pharynx, and epiglottis.

Finally, the olfactory neurons *are* the receptor cells and are uniquely exposed to the external environment, extending cilia along the epithelial surface, rather than being protected by epithelial and receptor cells as are gustatory neurons. Of course, receptor elements in both of these systems are subject to a constant barrage of chemical stimuli, some of which are potentially toxic, as well as being susceptible to direct injury from microbes. Although in both systems there is ongoing regeneration of these elements, in olfaction this process requires reinnervation of the olfactory bulb.

In short, it is not really surprising that smell dysfunction would be more common than taste dysfunction or that, in the words of Mackenzie (1884), smell might often be "altogether lost" (p. 323).

III. SMELL DISORDERS

A. Terminology

There are inconsistencies in the literature in the use of terms describing olfactory dysfunction. In general, *anosmia* is used to refer to an absence of smell function, and *hyposmia* to diminished smell sensitivity. Occasionally, however, the term *anosmia* seems to be used more broadly to encompass both of these conditions, and the term *specific anosmia* (deficit in olfactory sensitivity to a specific odorous compound or limited class of compounds, with intact general olfactory abilities) is often used to refer to relative as opposed to absolute insensitivity. A large number of specific anosmias have been described, and these hold considerable interest in terms of their implications for both the genetic involvement in olfaction and olfactory receptor mechanisms (see, e.g., Amoore, 1977); they do not, however, typically present as clinical problems and will not be considered further here. A final point that should be made in this context is that, even when used in its more restricted sense, anosmia does not necessarily imply a complete inability to detect the presence of a volatile odorous compound at any concentration, because most odorous compounds, at least at high concentrations, also stimulate nasal fibers of the trigeminal (V) nerve (Doty et al., 1978).

The terms *dysosmia* and *parosmia* are widely used to refer both to cases in which the patient experiences an odor sensation in the absence of an odorous stimulus and to those in which there are distortions in the perceived qualities of odorous stimuli. The former condition is often referred to more specifically as *phantosmia*. Although Henkin (1987) proposed an elaborate classification system that included three terms (cacosmia, heterosmia, and parosmia) to describe different forms of odor distortions, there is no generally agreed upon term to refer specifically to distortions. In the present review, therefore, use of the term dysosmia will be restricted to odor quality distortions, and phantosmia will be used to refer to spontaneous (unstimulated) odor sensations. [It should perhaps be noted that at least one modern author has used dysosmia to refer generally to any disruption in olfactory function that manifests in reduced ability to identify common odorants (Wright, 1987).]

B. Assessment

Both implicitly in his comment on the frequent confusion of taste with smell by medical writers (see Section II), and more explicitly in his criticism of a case report because the patient's *"sense of smell was never actually tested"* (p. 320), Mackenzie (1884) seemed to recognize that one of the weaknesses of the 19th century medical literature on chemosensory disorders was the lack of standardized assessment procedures to document both the nature and the degree of dysfunction. To a large extent, this problem has been remedied in recent years, at least in the case of olfactory assessment.

Researchers at MJC, UPenn, and the CCCRC have each developed measures of olfactory function that have been administered to large samples of healthy volunteers to establish normative values (Cain, Gent, Goodspeed, & Leonard, 1988; Cowart, 1989; Doty, Shaman, & Dann, 1984). In each case the test battery includes a measure of threshold sensitivity and a multiple-choice odor identification test. Both MJC and UPenn assess threshold sensitivity using phenyl ethyl alcohol (PEA), an odor compound that elicits little or no nasal trigeminal response at any concentration (Doty et al., 1978; Kobal, 1982), and administer 40-item, four-alternative, forced-choice identification tests [the test used at UPenn, the University of Pennsylvania Smell Identification Test (UPSIT), employs microencapsulated odors and is commercially available]. MJC also obtains a measure of threshold sensitivity to pyridine (Sherman, Amoore, & Weigel, 1979). The battery employed by the CCCRC differs in that threshold for butanol is assessed, and the identification measure is not forced-choice and is based on responses to only 7 items.

The similar rates of diagnosis of olfactory dysfunction in the three centers, noted in Section II, suggest that variations in the stimuli presented and procedural details may have little impact on the results of olfactory testing in a clinical setting, although the lack of a forced-choice format in their identification test may contribute to the somewhat higher rate of olfactory diagnosis reported by the CCCRC than by MJC and UPenn. Nonetheless, the composite score obtained from the CCCRC tests has been shown to correlate well with UPSIT scores alone (Smith, 1988). Moreover, all three centers have reported high correlations between their measures of threshold sensitivity and odor identification ability (Cain et al., 1988; Cowart, 1989; Doty, Shaman, & Dann, 1984), and it has been suggested that both types of test measure essentially the same property (sensitivity) and are redundant (Cain et al., 1988).

At MJC, however, we have observed that this correlation breaks down in individuals complaining of dysosmia (Cowart, Garrison, Young, & Lowry, 1989). For example, in our current sample of 797 patients who have completed both the PEA threshold and odor identification tests, the correlation between these two measures is 0.75 among patients with no complaint of odor quality distortions ($n = 621$), indicating that in this group the score on one test accounts for 56% of the variance on the other. Among patients who report distortions ($n = 176$), however, the correlation falls to 0.45, accounting for only 20% of the variance. The difference between these two correlations is statistically reliable ($z = 5.67$, $p < .0001$). The relatively poor correlation between threshold and identification measures in patients reporting dysosmia reflects the fact that most of these patients evidence reasonably good absolute sensitivity to the presence of odorous stimuli but have substantial difficulty identifying those stimuli, presumably due to the quality distortions they experience (see Zilstorff & Herbild, 1979). At MJC, we now use a discrepancy in performance on threshold and identification measures, in conjunction with patient report of distortions, to assign a primary olfactory diagnosis of dysosmia (as opposed to hyposmia or anosmia), and of those patients with measurable smell dysfunction, 14.1% (78/554; 9.4% of the total sample) meet these criteria for dysosmia. As described in Section III,D, there is preliminary evidence that this diagnostic distinction may have prognostic significance, at least in terms of patients' subjective reports.

No performance measure has been found to distinguish phantosmia, and all of the clinical centers continue to rely exclusively on patient report of this form of dysfunction. Most patients who complain of odor phantoms do, however, evidence measurable olfactory dysfunction; for example, of 128 such patients seen at MJC,

73.4% produced aberrant scores on olfactory testing, with 41 being diagnosed as hyposmic, 33 as anosmic, and 20 as primarily dysosmic.

Finally, Mackenzie (1884) suggested that when patients present with a complaint of smell impairment, function should be tested separately in each nostril to determine "whether the sense is destroyed on one side, or blunted on both" (p. 325), although his rationale for this is questionable ["in loss of smell dependent on injury to the seventh or fifth nerves the affection is almost always unilateral" (p. 325)]. Both UPenn and the CCCRC routinely administer either a threshold test or both threshold and identification tests unilaterally, but neither has reported the frequency with which unilateral olfactory dysfunction is observed, or its etiologic/prognostic significance. Thus, the utility of routine unilateral clinical assessment of olfaction has yet to be established. On the other hand, Leopold has proposed, based on clinical experience, that phantosmias and, perhaps, dysosmias that arise from peripheral damage to the olfactory epithelium may almost always be unilateral (D. A. Leopold, personal communication, May, 1993). Because this could have implications for treatment, or at least for recommendations to aid patients in coping with their dysfunction, it deserves further study and documentation.

C. Etiologies

Mackenzie (1884) suggested that the "most common cause of anosmia is prolonged catarrh" (p. 321), with the term *catarrh* being used to refer to inflammation of the nasal membrane with associated changes in mucous discharge (pp. 197–234) and, thus, to any form of rhinitis. Consistent with this observation, modern chemosensory clinical centers (Mott & Leopold, 1991; B. J. Cowart, I. M. Young, R. S. Feldman, & L. D. Lowry, unpublished data, 1996) have found diseases of the nose and/or paranasal sinuses to contribute to the plurality of cases of documented smell dysfunction (~30%; 15–29% of all presenting cases in the same centers). (It is, however, noteworthy that Mackenzie does not mention the paranasal sinuses, much less sinus disease, anywhere in his text.) Smell losses associated with nasal/sinus disease (NSD) can be profound; indeed, the majority of NSD patients who present to chemosensory clinical centers are found to be anosmic (Cain et al., 1988; Cowart et al., unpublished data, 1996). Reports of phantom odors, sometimes in association with odor quality distortions, are not uncommon in this patient group (Mott & Leopold, 1991); at MJC 15.2% of patients with NSD-related dysfunction ($n = 151$) present with such complaints (11.3% with phantoms alone)—but quality distortions by themselves are relatively rare (reported by fewer than 6% of these patients at MJC).

Mechanical obstruction of the access of molecules to the olfactory receptors in rhinitis (and/or nasal polyposis) would seem to provide an easy explanation for this form of loss, and although he mentions that as a potential contributing factor, Mackenzie (1884) seemed to favor, with no specific evidence, an explanation based on changes in the normal moisture of the olfactory neuroepithelium, and possible destruction of the receptor cell processes by the "inflammatory exudation" (p. 321). It now seems clear that pathologic mechanisms other than mechanical obstruction are involved in NSD-related smell loss. For example, obstruction of the nasal airway in patients with allergic rhinitis has not been found to be associated significantly with olfactory sensitivity (Cowart et al., 1993). Moreover, at least some patients with NSD and diminished olfactory function can be shown through modern endoscopic and computerized tomography (CT) scanning techniques to be free of significant obstruction, but they nonetheless recover olfaction with treatment of their NSD (Mott & Leopold, 1991). Although damage to the olfactory receptors by products of inflammation is theoretically possible, the fact that this form of loss often responds rapidly to the antiinflammatory effects of systemic corticosteroid therapy (Cain et al., 1988; Mott & Leopold, 1991; Jafek, Moran, Eller, Rowley III, & Jafek, 1987) suggests that is not the major underlying mechanism; in addition, the few available histologic studies of the olfactory mucosa in patients with NSD-related anosmia have found it to be essentially normal (Douek, Bannister, & Dodson, 1975; Yamagishi, Hasegawa, & Nakano, 1988). Alternatively, edema of the neuroepithelium could stretch the olfactory neurons and impede synaptic transmission (Mott & Leopold, 1991).

Finally, Mackenzie placed early, insightful emphasis on the critical importance to olfaction of the "normal moisture of the [neuroepithelial] surface" (p. 321). Changes in the composition of the mucous overlaying the olfactory receptor cilia could interfere with transport of odorant molecules to the receptors and/or with receptor binding. The possible role of such changes in NSD-related olfactory loss deserves further study.

Mackenzie (1884) seemed unaware of what now appears to be the second most common etiologic basis for smell dysfunction, prior upper respiratory infection (URI)—or at least he did not distinguish these dysfunctions from those secondary to ongoing inflammatory diseases of the nose and/or sinuses. The chemosensory clinical centers report that prior URI is implicated in 14–26% of all presenting cases (Mott & Leopold, 1991; Cowart et al., unpublished data, 1996). Because viral URI can precede and incite secondary bacterial sinusitis (Mott, 1991), patients with these two forms of smell dysfunction may, in fact, present with similar histories. A number of characteristics of URI-related smell dysfunction do, however, clearly differentiate it from NSD-related loss.

First, URI-related dysfunctions tend to occur less frequently in young individuals than do NSD-related losses (Leopold, Hornung, & Youngentob, 1991; Mott & Leopold, 1991). At MJC, for example, 12.3% of our patients with apparent URI-related dysfunction have been under 40 years of age (18 of 146 patients), whereas over 18% of those whose smell dysfunction appeared to be related to ongoing NSD have been that young (28 of 151 patients); this difference is not statistically significant, however. The post-URI patient group also includes proportionately more women than does the NSD group (MJC data: 68.5% vs. 51%, respectively; Fisher's Exact Test, $p = .003$) (see also Leopold et al., 1991; Mott & Leopold, 1991). In addition, URI-related dysfunctions are significantly less likely to manifest as anosmia than are NSD-related dysfunctions (MJC data: 20.5% vs. 58.9%, respectively; Fisher's Exact Test, $p < .0001$; see also Cain et al., 1988; Leopold et al., 1991; Mott & Leopold, 1991).

On the other hand, several authors have reported that dysosmia and phantosmia are frequent components of URI-related olfactory dysfunction (Scott, 1989; Zilstorff & Herbild, 1979), with approximately half of the patients reporting one or both of those symptoms (Henkin, Larson, & Powell, 1975; but see Leopold et al., 1991). At MJC, we have found dysosmia in particular to distinguish URI and NSD patients. Odor quality distortions with no phantom experience are reported by 32.2% of our URI patients (8.9% report phantoms alone and 15.8% report both), and significantly more URI and NSD patients receive a primary olfactory diagnosis of dysosmia (27.4% vs. 3.3%, respectively; Fisher's Exact Test, $p < .0001$).

Finally, and most basically, in URI-related dysfunction there appears to be damage to peripheral olfactory receptors (Douek, Bannister, & Dodson, 1975; Jafek, Hartman, et al., 1990; Yamagishi et al., 1988). In the largest histopathological study of postviral olfactory dysfunction to date, Jafek, Hartman, et al. (1990) observed varying degrees of olfactory epithelial destruction in these patients, ranging from virtually total destruction of olfactory receptor neurons to reductions in the number of receptors with patches of epithelium having a relatively normal appearance. Furthermore, the severity of histopathologic change was found to be correlated with the observed olfactory deficit. The authors speculate that either abnormal axonal reconnection of regenerating neurons or the patchy degeneration (and/or incomplete regeneration) they observed, which could alter the sorptive characteristics of the receptor sheet, might underlie the odor quality distortions often reported by URI patients.

The third major etiologic basis for smell dysfunction at chemosensory clinical centers is head trauma, accounting for 10–19% of all presenting cases (Mott & Leopold, 1991; Cowart et al., unpublished data, 1996). The first reports of posttraumatic anosmia appeared in the medical literature in the latter half of the 19th century (see Sumner, 1964), and Mackenzie (1884) noted that these types of cases "are by no means rare" (p. 322). Current estimates suggest that 20–30% of head trauma patients sustain some degree of olfactory impairment (anosmia, hyposmia, or dysosmia) (Costanzo & Zasler, 1991). The likelihood of posttraumatic olfactory loss appears to increase with the severity of the injury (Costanzo & Zasler, 1991), although it can occur after trivial injuries with no associated posttraumatic amnesia (Sumner, 1964), and blows to the occipital region may be most likely to produce smell dysfunction (Sumner, 1964). As is the case in NSD-related loss, the majority of trauma patients who present to chemosensory clinical centers are found to be anosmic (Cain et al., 1988; Cowart et al., unpublished data, 1996). Phantosmia and dysosmia are more common in trauma-related than in NSD-related dysfunctions, however; at MJC, 23.3% of trauma patients with smell dysfunction ($n = 86$) report distortions, 18.6% report a phantom smell, and 5.8% report both, with 16.3% meeting our diagnostic criteria for dysosmia.

Mackenzie (1884) postulated that the basis for olfactory loss in trauma cases was a separation of the olfactory bulbs from the brain. Although damage to higher neural pathways is possible, it is now believed that the most common mechanism is tearing or severing of the olfactory neuron axons at the point at which they pass through the small openings in the cribriform plate, as a result of the coup contra coup forces associated with head injury (Costanzo & Zasler, 1991). Jafek, Eller, Esses, & Moran (1989) have reported that observations from the ultrastructural examination of olfactory epithelium biopsies from five patients with posttraumatic anosmia seem to be consistent with traumatic severing of the olfactory filaments at the cribriform plate, followed by regeneration of the neuroepithelium and a failure of regenerating axons to reach the olfactory bulb (possibly as a result of fibrotic healing of the lamina cribrosa of the cribriform plate and closure with scar tissue). Specifically, they consistently observed disruption in the epithelial organization consonant with a regenerating epithelium, large numbers of axon tangles throughout the epithelium, and few, if any, olfactory cilia projecting from the receptor cells, a condition that they postulated reflects the dependency of olfactory dendrite ciliogenesis on axonal contact with central nervous system tissue (but see the following discussion of Kallmann syndrome).

Mackenzie (1884) also specifically mentioned aging, exposure to "irritant vapors" (p. 321), and heredity or "congenital deficiency of the olfactory nerves" (p. 323) as etiologic factors in smell dysfunction. Together with the three major factors discussed previously, these probably encompass most instances of olfactory dysfunction in which a causal condition can be identified, although a number of medical conditions (perhaps most notably Alzheimer disease) and some medications have also been associated with smell disorders (for overviews of this literature, see Mott & Leopold, 1991; Schiffman, 1983a, 1991).

As noted in Section I, there is now extensive documentation of age-related decline in olfactory sensitivity, even in the healthy elderly. This form of loss apparently occurs gradually, is not typically complete, and often seems to go unnoticed by individuals experiencing it (Cowart, 1989; Nordin, Monsch, & Murphy, 1995; Stevens & Cain, 1985). Nonetheless, it may be of magnitude sufficient to render them vulnerable to chemical hazards such as gas leaks (Chalke & Dewhurst, 1957; Stevens, Cain, & Weinstein, 1987; Wysocki & Gilbert, 1989) and to impact on food flavor perception (Cain, Reid, & Stevens, 1990; Murphy, 1985; Schiffman, 1977, 1979; Stevens & Cain, 1986).

Given the relatively unprotected position of olfactory receptor neurons, it is not surprising they are susceptible to damage from pollutants in the ambient air. There are substantial animal toxicological data demonstrating damage to the olfactory neuroepithelium and bulb by airborne chemicals (e.g., Keenan, Kelly, & Bogdanffy, 1990; Min, Rhee, Choo, Song, & Hong, 1994; Nikula & Lewis, 1994, and see

Schwartz, Doty, Monroe, Frye, & Barker, 1989), as well as a large but scattered literature on the adverse effects on the sense of smell of occupational exposures to industrial chemicals (see Amoore, 1986). In humans, both acute and chronic exposures to a variety of chemical agents have been associated with olfactory dysfunction, which may be either temporary or permanent (Amoore, 1986). This factor may, in fact, play an important role in age-related smell loss.

There is still relatively little known about genetic/congenital olfactory dysfunction. The principal genetic syndrome associated with anosmia is Kallmann syndrome, which is also characterized by hypogonadotropic hypogonadism. These symptoms appear to be secondary to a failure during embryonic development of gonadotropin-releasing hormone-producing neurons to migrate to the brain from the olfactory placode (Rugarli & Ballabio, 1993), and to insufficient or absent synaptic connections between olfactory neurons and cells in the olfactory bulb (Truwit, Barkovich, Grumbach, & Martini, 1993), which is either aplastic or hypoplastic (Bajaj et al., 1993; Klingmüller, Dewes, Krahe, Brecht, & Schweikert, 1987; Knorr, Ragland, Brown, & Gelber, 1993; Yousem, Turner, Li, Snyder, & Doty, 1993). On the basis of histopathological studies of biopsy specimens from the olfactory region, Jafek, Gordon, Moran, and Eller (1990) reported an apparently complete absence of olfactory epithelium in one patient with Kallmann syndrome, as well as in six other cases of congenital anosmia with varying medical and family histories. In a similar study of one anosmic Kallmann's patient, however, Schwob, Leopold, Szumowski, and Emko (1993) found olfactory epithelium similar to that observed in bulbectomized animals, with olfactory neurons that appeared to be structurally immature. More recently, Rawson et al. (1995) demonstrated functional maturity (odorant-specific responsiveness) in isolated olfactory neurons from each of two anosmic patients with Kallmann syndrome. Thus, anosmia in these patients was not due to a lack of functioning olfactory neurons, and differentiation of these neurons may not require contact with the olfactory bulb.

There are scattered reports of other forms of familial anosmia (e.g., Singh, Grewal, & Austin, 1970); at MJC, however, fewer than 40% of the patients we have seen who did not recall ever being able to smell (11 of 29 patients) were aware of any family history of a similar problem. It is impossible to rule out early childhood loss secondary to a head injury or upper respiratory infection in these cases. Nonetheless, it is somewhat surprising that 3–4% of the patients presenting to chemosensory clinical centers report lifelong anosmia (Deems et al., 1991; Cowart et al., unpublished data, 1996), suggesting that, although not common, this condition may not be extremely rare.

Finally, Mackenzie (1884) admitted "there are some cases of anosmia in which it is impossible to discover any cause for the loss" (p. 323). This is unfortunately still true. Various centers for chemosensory evaluation have reported that no causative condition for chemosensory dysfunction can be identified in 10–24% of their patients (Mott & Leopold, 1991), although the higher estimates include at least a small proportion of individuals whose dysfunction is limited to taste, for which our understanding of causal factors is even poorer than is the case with smell (see Section IV, C).

D. Prognosis

The potent antibiotic and antiinflammatory agents now commonly used in the medical treatment of NSD were not available in the 19th century, and as a result, Mackenzie (1884) was not altogether optimistic about the prognosis of NSD-related smell loss, although he did recognize it to be more favorable than in cases of traumatic lesion. In fact, this form of loss is now, to a large extent, defined by the fact that it reverses on treatment (Mott & Leopold, 1991). Long-term management is, however, complicated by the chronicity of the underlying disorders, and even patients who obtain effective treatment may be subject to recurring episodes of loss.

In cases of smell dysfunction secondary to peripheral nerve damage, such as is presumably the case in most instances of disorders that follow URI or toxin exposure, and in at least some cases of trauma-related disorder, gradual spontaneous recovery is theoretically possible due to the regenerative capacity of the olfactory neural receptors. Neither the time course of recovery nor factors affecting the likelihood of recovery have, however, been fully elucidated.

The most extensive prognostic studies have been conducted in patients suffering from posttraumatic smell dysfunction. Estimates of the incidence of full or partial recovery vary widely. In the two largest of these studies, however, Costanzo and Becker (1986) and Sumner (1964) both reported improvement in 30–40% of the patients examined. Data from both studies also suggest there is a decrease in the rate of recovery after 3–6 months, and thus, possibly more than one mechanism underlying the recovery process. Sumner (1964) speculated that rapid early recovery of olfaction may reflect the resolution of edema or blood clots, whereas the slower recovery rates observed at longer durations may reflect the regeneration and replacement of damaged olfactory neurons.

Data on spontaneous recovery in URI-related olfactory disorders are limited and conflicting. Mott and Leopold (1991) reported a longitudinal study of 40 post-URI patients at the CCCRC, in which 15% showed improvement in olfactory test scores of 40% or more after an average of 26 months, even though initial testing was performed, on average, 2 years after the onset of the problem. In a similar study of 35 patients at the Olfactory Referral Center of the State University of New York (SUNY) Health Science Center, however, only 1 patient showed improvement of more than 15% on an odor identification test (Mott & Leopold, 1991). Similarly, Deems et al. (1991) reported no change in the mean odor identification scores (on the UPSIT) of an unspecified number of URI patients retested after an interval of 5 months to 6.4 years. On the other hand, in a recent study of 21 URI patients, Duncan and Seiden (1995) reported that 67% (14) improved their UPSIT score by 4 points or more after an average of 3 years, with 13 of these patients also reporting subjective improvement.

There is also some debate with regard to the prognostic significance of the development of dysosmic symptoms in head trauma and URI patients. It has been speculated that odor quality distortions could reflect either degenerative or regenerative changes in the olfactory epithelium. Leigh (1943) reported that in 72 cases of olfactory impairment following head injury, 3 of the 12 patients complaining of parosmia (25%) noted eventual recovery, whereas only 3 of 60 with complaints of simple loss (5%) reported recovery. Retest data obtained by Deems et al. (1991) and Duncan and Seiden (1995) do not, however, support the hypothesis that recovery is more likely in patients with dysosmia, although the data also do not suggest these patients are more likely to evidence a decline in function.

In both of the latter studies, patients reporting phantom smells, as well as those whose test scores indicated they had no residual olfactory function even though they reported phantoms and/or distortions, were included in the "dysosmic" category. As described in Section III,B, at MJC patients are considered primarily dysosmic only if they evidence odor quality distortions in the absence of substantial loss in absolute olfactory sensitivity. On average, the odor identification performance of these patients is no better than that of patients diagnosed as hyposmic, although (by definition) their threshold sensitivity is greater. In follow–up interviews conducted with 268 patients (109 hyposmics, 115 anosmics, and 44 dysosmics), significantly more dysosmics (61.3%) than either anosmics (19.1%) or hyposmics (36.7%) reported having experienced improvement in smell function since their evaluation ($\chi^2 = 26.6$; $p < .0001$). Of course, subjective reports of improvement could reflect either a real change in olfactory function or the patient's having adapted to the problem, and firm conclusions must await objective testing in these patient groups.

No effective treatment for smell disorders other than those associated with nasal/sinus disease has been identified. Zinc is often prescribed, but has been shown to be no more effective than placebo in a double-blind study (Henkin, Schechter, Friedewald, Demets, & Raff, 1976). A number of other treatments have been suggested, especially for problems secondary to URI, but controlled clinical trials have not been conducted. Perhaps the most intriguing of these is vitamin A therapy. In a reasonably large study, although not one that was blinded or included a untreated control group, Duncan and Briggs (1962) reported that 50 of 56 anosmic patients treated with intramuscular injections of vitamin A alcohol in oil regained full or partial olfactory function. The plurality of these patients had suffered a URI-related loss; neither the one patient with a congenital loss nor the three with losses secondary to head trauma responded to this treatment. More recently, Roydhouse (1988) also reported responsiveness to oral retinoid therapy in two patients, one whose dysfunction was secondary to URI, the other whose etiology was unknown. Because vitamin A is present in the olfactory epithelium (Duncan & Briggs, 1962) and may play a role in neural regeneration, there is some rationale for this therapy and further research into its efficacy would seem to be warranted.

IV. TASTE DISORDERS

A. Terminology

The terms commonly applied to taste dysfunction parallel those for smell disorders, and there is somewhat greater consistency in their usage; however, they fail to make some distinctions that probably should be made. As might be expected, *ageusia* refers to a complete absence of gustatory function and *hypogeusia* to diminished taste sensitivity. The difficulty here is that in the gustatory system fundamentally different transduction sequences underlie the perception of different taste qualities, and probably, at least in the case of bitter, that of different compounds with the same quality (see Ch. 1). Thus, an individual could potentially experience either a total but isolated loss in sensitivity to compounds eliciting a specific quality, or a more generalized diminution in sensitivity to compounds eliciting a variety of taste qualities, and in both cases be classified simply as hypogeusic. Some instances of specific gustatory insensitivity, such as genetic insensitivity to the bitter taste of phenylthiocarbamide (PTC) and its structural analogs (see Kalmus, 1971), are (like specific anosmias) unlikely to present as clinical problems; a specific loss in sensitivity to sweets might well, however.

As is also the case in smell loss (Section III,A), it should be noted that even a completely ageusic patient will be able to detect high concentrations of some gustatory stimuli through stimulation of fibers of the trigeminal (V) nerve. For example, both sodium chloride (salt) and sour acids can elicit oral irritation/trigeminal responses (e.g., Green & Gelhard, 1989; Bryant & Moore, 1995). Whether sweet and bitter compounds do as well is unclear, although neither of the two ageusic patients who have been evaluated at MJC consistently responded to high concentrations of either sucrose or quinine sulfate (markedly elevated, but stable, thresholds for salt and citric acid were obtained in both cases).

For the most part, the term *dysgeusia* is used in the clinical taste literature essentially as a synonym for *phantogeusia,* to refer to the experience of a taste sensation in the apparent absence of a gustatory stimulus. Primary distortions in the perceived qualities of gustatory stimuli are, unlike olfactory distortions, not well documented in clinical settings, although they may occur. Phantom tastes are sometimes accompanied by oral burning sensations, or such symptoms may appear alone in what is often referred to as *burning mouth syndrome.* This symptom complex will not be considered here, but a concise and balanced review of the relevant clinical literature may be found in Forman and Settle (1990a, 1990b).

B. Assessment

Clinical assessment of taste is not as well developed or standardized as that of olfaction. Because of the possibility of clinically significant, quality-specific loss, it is probably important to obtain some measure of responsivity to representatives of the four generally agreed-upon taste qualities (sweet, salty, sour, bitter). The most common choices are sucrose, sodium chloride, citric or hydrochloric acid, and quinine (sulfate or hydrochloride) or caffeine. This, obviously, creates logistical difficulties, especially given the necessity of preparing fresh taste stimuli frequently. An alternative to the use of chemical stimuli is measurement of electric taste via electrogustometry (see Frank & Smith, 1991). Although not sensitive to quality-specific loss, this form of assessment offers advantages not only in terms of simplicity and portability, but also in the degree of precision with which the stimulus may be applied. It has not, however, been widely used in the United States, and none of the chemosensory clinical research centers here relies exclusively, or even primarily, on electrogustometric assessment in taste evaluations.

Gent, Frank, and Mott (1997) reviewed and provided detailed descriptions of many of the taste assessment procedures used in clinical practice. Of the three centers on which this review has focused, MJC and UPenn both rely primarily on measures of whole-mouth threshold sensitivity to the basic tastes in the diagnosis of taste loss (although UPenn does not include a bitter stimulus in this assessment). In addition, both utilize suprathreshold quality identification and category scaling of taste intensity as screening measures and to supplement the interpretation of threshold results. [Although quality identification has proved to be a very useful clinical tool in olfaction, its utility in gustatory assessment is limited by the fact that taste quality confusions (particularly sour–bitter, but also sour–salty and salty–bitter) are not uncommon in the general population.] As indicated (Section II), both MJC and UPenn also report very low rates of diagnosis of taste loss. The CCCRC, on the other hand, does not assess threshold sensitivity, but relies on a suprathreshold intensity scaling task that employs the psychophysical technique of magnitude matching to allow for the direct comparison of intensity ratings across subjects (see Bartoshuk, Gent, Catalanotto, & Goodspeed, 1983). Presumably, this marked difference in assessment procedures accounts for the substantially higher rate of taste diagnosis that has been reported at the CCCRC (see Section II).

In the context of the potential for quality-specific loss in taste, it is interesting to note that, in olfaction, not only are threshold measures highly correlated with identification measures (see Section III,B), but at MJC we have found that the two olfactory thresholds we obtain (for odors that are qualitatively very different) are also highly correlated. In a sample of 740 patients who completed both odor threshold tests as well as all four measures of taste threshold, the correlation between the odor thresholds is .67; in contrast, correlations among taste thresholds are significantly lower, ranging from .25 (for sucrose and quinine) to .46 (for salt and citric acid) ($p < .0001$ in all comparisons with the odor threshold correlation). Indeed, among those patients considered to evidence taste loss on the basis of their threshold test results, almost half (36 of 73) showed clinically significant elevation in only a single taste threshold (in 19 of these cases, salt sensitivity was affected), and in only 6 cases (8.2% of those with any loss) were thresholds for all four tastes elevated.

As is the case with phantosmia, no performance measure clearly distinguishes phantogeusia. At MJC, we have found that only about 18% of the patients complaining of taste phantoms evidence measurable, whole-mouth taste loss, which is substantially lower than the frequency with which measurable smell dysfunction is observed in patients with phantom smell complaints (~73%; see Section III,B). Nonetheless, diminished taste sensitivity is significantly more common among patients reporting taste phantoms than among those who do not (only 7% of whom evidence taste loss; Fisher's Exact Test, $p = .0015$).

Finally, because the gustatory system is bilaterally innervated by three cranial nerves, there is the possibility of loss localized to one or more receptor fields as a result of peripheral or central lesions. MJC, UPenn, and the CCCRC all employ regional gustatory testing to supplement whole-mouth testing in some patients. In all clinics, suprathreshold concentrations of chemical stimuli are swabbed or pipetted onto discrete regions of the tongue (and, when swabs are used, the palate), and patients rate the intensity of the resulting taste and label its quality; left–right and anterior-posterior differences in responsiveness are then examined. UPenn also utilizes electrogustometry in regional testing, which makes assessment of left–right thresholds feasible. In fact, regional losses often are not associated with any subjective change in taste experience; they may, however, underlie some instances of taste phantoms (see Section IV,C).

C. Etiologies

Detailed reports of etiologic factors contributing to taste dysfunction in patients presenting to chemosensory clinics are largely lacking. Thus, the most common causes of such dysfunctions cannot be identified with any confidence. Indeed, at MJC, no clear precipitating event or readily identifiable underlying pathology has been evident in over 60% of the cases of taste loss and/or phantoms we have evaluated.

Based on the sheer number of clinical reports in the literature (see Mott & Leopold, 1991; Rollin, 1978; Schiffman, 1983a, 1991), one might argue that the single most common etiologic factor contributing to taste dysfunction is probably medication usage. One summary report from a Japanese taste clinic is consistent with this (Tomita, 1990). In the United States, however, cases of drug-related taste disturbances seem to be underrepresented among clinical research center patients, possibly because of their widespread recognition by medical practitioners and patients' ability to tie changes in taste to the use of specific medications. Different mechanisms undoubtedly underlie the gustatory effects of different medications. For example, taste perception could be affected through the alteration of salivary constituents, through vascular tastes, through disruption of transduction/receptor mechanisms, and through alterations in the central processing of gustatory input. In very few cases, however, have the specific mechanisms underlying taste effects of a given medication been elucidated. A better understanding of these effects might shed light on both normal gustatory function and the vulnerabilities of this system.

Oral health problems in the form of poor oral hygiene or periodontal disease are obvious potential sources of phantogeusias. Xerostomia, whatever its etiology, might also be expected to be associated with phantogeusia and/or taste loss as a result of its adverse impact on oral clearance and on the integrity of the teeth and oral mucosa; however, even severe, chronic failure of all salivary glands does not necessarily lead to taste complaints or abnormalities in taste function (Weiffenbach, Fox, & Baum, 1986; Weiffenbach, Schwartz, Atkinson, & Fox, 1995), a finding that attests to the remarkable resilience of the taste receptors.

Xerostomia is also one of several factors, including the use of dentures, antibiotics, or corticosteroids and immunological deficiencies, that can predispose to the overgrowth of oral *Candida* (Forman & Settle, 1990a). Brightman, Guggenheimer, and Ship (1968) found that subclinical elevations in oral *Candida* (i.e., without clinically evident thrush or angular cheilitis) produced bad tastes and burning sensations in a third of subjects in whom fungal overgrowth was induced by an oral tetracycline rinse. At MJC, we now routinely perform oral yeast cultures in patients complaining of phantogeusia. To date, just under a third (14 of 44) have been positive, and of the 7 patients we know to have received antifungal therapy, 5 reported resolution of their phantom. Similarly, Osaki, Ohshima, Tomita, Matsugi, and Nomura (1996) reported oral candidiasis to underlie dysgeusia in 3 of 14 patients they evaluated. Interestingly, these investigators also found candidiasis to contribute to simple taste loss in nearly a quarter of 25 hypogeusic patients.

Two of the most common causes of smell dysfunction, URI and head trauma, may also be associated with true taste problems, with both phantogeusias and losses having been reported (Costanzo & Becker, 1986; Costanzo & Zasler, 1991; Henkin, Larson, & Powell, 1975; Leopold et al., 1991). In both cases, taste symptoms are much less common than smell symptoms, and the underlying pathophysiology is not well understood. At MJC, about a fifth of our patients with taste loss have linked that symptom to a prior URI or head injury; almost all of these individuals also evidenced olfactory dysfunction. A smaller proportion (5%) of our phantogeusic patients (with no measured taste loss) have indicated their symptom was precipitated by either of these factors.

Two common surgical procedures may result in damage to the chorda tympani, which mediates taste perception on the anterior tongue. This nerve is frequently severed, stretched, or crushed during middle ear surgery (Bull, 1965; Chilla, Nicklatsch, & Arglebe, 1982; Grant, Miller, Simpson, Lamey, & Bone, 1989), and chorda–lingual damage has been estimated to occur in as many as 11% of patients who undergo third-molar extractions (Blackburn & Bramley, 1989). In cases of bilateral chorda tympani section, patients are often aware of some diminution in taste function, but they only rarely report a loss following unilateral damage (Bull, 1965; Grant et al., 1989). On the other hand, complaints of phantogeusia following surgical damage to the chorda appear to be common (Bull, 1965; Moon & Pullen, 1963). Taste phantoms may also be experimentally induced by anesthetization of the chorda tympani (Yanagisawa, Bartoshuk, Karrer, Kveton, & Catalanotto, 1992).

Disinhibition of responses from taste receptors innervated by the glossopharyngeal nerve has been proposed as a mechanism to explain both the limited impact of chorda damage on whole-mouth taste perception and the occurrence of phantoms (Catalanotto, Bartoshuk, Östrom, Gent, & Fast, 1993; Kveton & Bartoshuk, 1994; Yanagisawa et al., 1992). A decrease in spontaneous activity at the level of the nucleus of the solitary tract has also been suggested as a mechanism for these phenomena (Dinkins & Travers, 1996). Finally, abnormal functioning of the damaged nerve at the periphery may underlie phantoms in some cases (Bartoshuk, Kveton, & Lehman, 1992).

Since publication of an apparently successful, single-blind trial of the efficacy of zinc supplementation in reversing hypogeusia (Schechter, Friedewald, Bronzert, Raff, & Henkin, 1972), zinc deficiency has received considerable attention as a potential etiology for taste dysfunction, even though a subsequent double-blind trial failed to show any significant difference between the effects of zinc and placebo (Henkin et al., 1976). Some controlled studies of documented zinc deficiency in specific disease states do indicate it may be associated with taste loss that reverses on treatment with zinc (Atkin-Thor, Goddard, O'Nion, Stephen, & Kolff, 1978; Mahajan et al., 1980; Weisman, Christensen, & Dreyer, 1979), although the mechanisms by which zinc affects gustatory function are not known. Nonetheless, it seems unlikely zinc deficiency underlies most, or even many, cases of hypogeusia. The results of the double-blind study by Henkin et al. (1976) are consistent with this, as is the report by Deems et al. (1991) of no difference between the taste scores of patients presenting to UPenn who were taking zinc and those who were not. It should perhaps be noted, however, that Tomita (1990) has reported that the majority of a large sample of patients presenting to a Japanese clinic with taste complaints evidenced zinc deficiency and/or responded to zinc therapy. On the other hand, Osaki et al. (1996) found no evidence of zinc deficiency in their sample of Japanese patients with complaints of hypogeusia or dysgeusia (they did identify iron deficiency as a source of hypogeusia in 7 of 25 patients).

Finally, aging itself may be associated with diminished taste sensitivity. In the healthy elderly, this loss is, on average, less pronounced than are declines in olfactory sensitivity (Cowart, 1989; Stevens, Bartoshuk, & Cain, 1984), and may be quality or compound specific, at least in terms of the degree of change (Cowart, 1981, 1989; Cowart, Yokomukai, & Beauchamp, 1994; Murphy & Gilmore, 1989; Schiffman et al., 1994; Stevens, 1996; Weiffenbach, Baum, & Burghauser, 1982). Thus, patients with simple age-related taste losses are probably even less likely than are those with age-related smell losses to present with a complaint in a clinical setting. Aging or factors associated with aging may, however, render individuals more vulnerable to taste dysfunctions that do lead them to seek medical assistance. For example, among patients presenting to MJC, the elderly ($\geq$65 years of age) are significantly more likely to complain of phantogeusia than are the middle aged (45–64 years) and young (<45 years); the respective percentages of patients in each group with this complaint are 28.6, 14.9, and 12.6% ($\chi^2 = 24.7$; $p < .0001$). A similar relationship is not seen in complaints of phantosmia. Elderly patients are also more likely than middle-aged and young patients to evidence diminished taste sensitivity; taste loss is diagnosed in 13.2, 9.8, and 4.5% of each group, respectively ($\chi^2 = 11.5$; $p < .005$). Again, we do not see this relationship in the diagnosis of smell loss in clinic patients.

D. Prognosis

Of the identifiable etiologies associated with taste dysfunction, several are amenable to intervention. The type or dosage of medications a patient is receiving can often be altered, and in most cases, the prognosis for drug-related taste problems is excellent (Rollin, 1978). Both periodontal disease and oral candidiasis respond to therapy, although these problems may often recur in susceptible patients. Xerostomia is difficult to manage, but recent clinical studies have shown oral pilocarpine to be beneficial in a variety of forms of xerostomia (Fox et al., 1991; Johnson et al., 1993; Rhodus & Schuh, 1991), and it has now been approved for this use.

As is the case with smell dysfunctions secondary to URI or head trauma, some proportion of taste dysfunctions with these etiologies evidence spontaneous recovery, although there are even fewer long-term, follow-up data available on taste than there are on smell. There are, however, indications that, at least in the case of trauma-related chemosensory dysfunction, taste is more likely than smell to recover (Sumner, 1967).

Spontaneous resolution of symptoms following damage to the chorda tympani nerve has also been reported. The Bull report (1965) suggests that taste phantoms typically last 3–4 months, although they may persist for more than a year in a minority of patients. Both Bull and Chilla et al. (1982) indicate that taste loss in the affected taste field (of which, as noted, patients are rarely aware) never recovers following transection of the nerve during middle ear surgery, but almost always does when the nerve is only stretched. More recently, Zuniga, Chen, and Miller (1994), in a preliminary report, demonstrated recovery of taste sensitivity in one patient following surgical intervention and repair of chorda–lingual damage sustained during a third-molar extraction.

Finally, as has been indicated (Section IV,C), many if not most cases of taste dysfunction are idiopathic in origin. The prognosis for these does not appear to be good. Among such patients who have been followed at MJC after a period of at least 6 months, 39% of those complaining of a phantom taste ($n = 41$) did report full or partial resolution of this symptom; however, almost 15% reported worsening and the remainder indicated there had been no change. In 15 cases of simple taste loss, only one patient reported any improvement in taste sensitivity.

V. CONCLUSION

A surprising amount was known about olfactory dysfunction over a century ago. Nonetheless, substantial progress has been made in the characterization of these disorders and in the treatment of those that are secondary to nasal/sinus pathology. A better understanding of the mechanisms underlying olfactory dysfunctions has also been gained, although there is still much to be learned in this area. Finally, we are still in the position of being unable to intervene in most cases of olfactory disruption. Increases in basic knowledge about the regenerative process in the olfactory epithelium may ultimately provide therapeutic direction.

Gustatory disorders remain relatively more obscure. Although they are also relatively less common, to the extent that their prevalence increases with age, we may expect a growing number of such cases. The development of standardized assessment techniques that can reasonably be applied in clinical settings is of critical importance in furthering the study of taste dysfunction.

Acknowledgments

This work was supported in part by NIH Grant P50 DC 00214. The authors thank Elizabeth Varga for her careful supervision of sensory testing and data management at the Monell–Jefferson Taste and Smell Clinic.

References

Amoore, J. E. (1977). Specific anosmia and the concept of primary odors. *Chemical Senses and Flavour, 2,* 267–281.

Amoore, J. E. (1986). Effects of chemical exposure on olfaction in humans. In C. S. Barrow (Ed.), *Toxicology of the nasal passages* (pp. 155–190). Washington, DC: Hemisphere.

Atkin-Thor, E., Goddard, B. W., O'Nion, J., Stephen, R. L., & Kolff, W. J. (1978). Hypogeusia and zinc depletion in chronic dialysis patients. *American Journal of Clinical Nutrition, 31,* 1948–1951.

Bajaj, S., Ammini, A. C., Marwaha, R., Gulati, P., Khetarpal, K., & Mahajan, H. (1993). Magnetic resonance imaging of the brain in idiopathic hypogonadotropic hypogonadism. *Clinical Radiology, 48,* 122–124.

Bartoshuk, L. M., Gent, J. F., Catalanotto, F. A., & Goodspeed, R. B. (1983). Clinical evaluation of taste. *American Journal of Otolaryngology, 4,* 257–260.

Bartoshuk, L. M., Kveton, J., & Lehman, C. (1992). Peripheral source of taste phantom (i.e., dysgeusia) demonstrated by topical anesthesia. [Abstract]. *Chemical Senses, 16,* 499–500.

Blackburn, C. W., & Bramley, P. A. (1989). Lingual nerve damage associated with the removal of lower third molars. *British Dental Journal, 167,* 103–107.

Brightman, V. J., Guggenheimer, J., & Ship, I. (1968). Changes in the oral microbial flora during treatment of recurrent aphthous ulcers. [Abstract]. *Journal of Dental Research, 47*(Suppl.), 126.

Bryant, B. P., & Moore, P. A. (1995). Factors affecting the sensitivity of the lingual trigeminal nerve to acids. *American Journal of Physiology, 268,* R58–R65.

Bull, T. R. (1965). Taste and the chorda tympani. *The Journal of Laryngology and Otology, 79,* 479–493.

Cain, W. S., Gent, J. F., Goodspeed, R. B., & Leonard, G. (1988). Evaluation of olfactory dysfunction in the Connecticut Chemosensory Clinical Research Center. *Laryngoscope, 98,* 83–88.

Cain, W. S., Reid, F., & Stevens, J. C. (1990). Missing ingredients: Aging and the discrimination of flavor. *Journal of Nutrition for the Elderly, 9,* 3–15.

Cain, W. S., & Stevens, J. C. (1989). Uniformity of olfactory loss in aging. *Annals of the New York Academy of Sciences, 561,* 29–38.

Catalanotto, F. A., Bartoshuk, L. M., Östrom, K. M., Gent, J. F., & Fast, K. (1993). Effects of anesthesia of the facial nerve on taste. *Chemical Senses, 18,* 461–470.

Chalke, H. D., & Dewhurst, J. R. (1957). Accidental coal-gas poisoning: Loss of sense of smell as a possible contributory factor with old people. *British Medical Journal, 2,* 915–917.

Chilla, R., Nicklatsch, J., & Arglebe, C. (1982). Late sequelae of iatrogenic damage to chorda tympani nerve. *Acta Otolaryngologica, 94,* 461–465.

Costanzo, R. M., & Becker, D. P. (1986). Smell and taste disorders in head injury and neurosurgery patients. In H. L. Meiselman & R. S. Rivlin (Eds.), *Clinical measurement of taste and smell* (pp. 565–578). New York: Macmillan.

Costanzo, R. M., & Zasler, N. D. (1991). Head trauma. In T. V. Getchell, R. L. Doty, L. M. Bartoshuk, & J. B. Snow, Jr. (Eds.), *Smell and taste in health and disease* (pp. 711–730). New York: Raven.

Cowart, B. J. (1981). Development of taste perception in humans: Sensitivity and preference throughout the life span. *Psychological Bulletin, 90,* 43–73.

Cowart, B. J. (1989). Relationships between taste and smell across the adult life span. *Annals of the New York Academy of Sciences, 561,* 39–55.

Cowart, B. J., Flynn-Rodden, K., McGeady, S. J., & Lowry, L. D. (1993). Hyposmia in allergic rhinitis. *Journal of Allergy and Clinical Immunology, 91,* 747–751.

Cowart, B. J., Garrison, B., Young, I. M., & Lowry, L. D. (1989). A discrepancy between odor thresholds and identification in dysosmia. [Abstract]. *Chemical Senses, 14,* 692–692.

Cowart, B. J., Yokomukai, Y., & Beauchamp, G. K. (1994). Bitter taste in aging: Compound-specific decline in sensitivity. *Physiology & Behavior, 56,* 1237–1241.

Deems, D. A., Doty, R. L., Settle, R. G., Moore-Gillon, V., Shaman, P., Mester, A. F., Kimmelman, C. P., Brightman, V. J., & Snow, J. B., Jr. (1991). Smell and taste disorders. A study of 750 patients from the University of Pennsylvania Smell and Taste Center. *Archives of Otolaryngology. Head and Neck Surgery, 117,* 519–528.

Dinkins, M. E., & Travers, S. P. (1996). Alternative mechanism for taste compensation following chorda tympani anesthetization. [Abstract]. *Chemical Senses, 21,* 595–596.

Doty, R. L., Brugger, W. E., Jurs, P. C., Orndorff, M. A., Snyder, P. J., & Lowry, L. D. (1978). Intranasal trigeminal stimulation from odorous volatiles: Psychometric responses from anosmic and normal humans. *Physiology & Behavior, 20,* 175–185.

Doty, R. L., Shaman, P., Applebaum, S. L., Giberson, R., Siksorski, L., & Rosenberg, L. (1984). Smell identification ability: Changes with age. *Science, 22,* 1441–1443.

Doty, R. L., Shaman, P., & Dann, M. (1984). Development of the University of Pennsylvania smell identification test: A standardized microencapsulated test of olfactory function. *Physiology & Behavior, 32,* 489–502.

Douek, E., Bannister, L. H., & Dodson, H. C. (1975). Recent advances in the pathology of olfaction. *Proceedings of the Royal Society of Medicine, 68,* 467–470.

Duncan, H. J., & Seiden, A. M. (1995). Long-term follow-up of olfactory loss secondary to head trauma and upper respiratory tract infection. *Archives of Otolaryngology, Head and Neck Surgery, 121,* 1183–1187.

Duncan, R. B., & Briggs, M. (1962). Treatment of uncomplicated anosmia by vitamin A. *Archives of Otolaryngology, 75,* 116–124.

Estrem, S. A., & Renner, G. (1987). Disorders of smell and taste. *Otolaryngologic Clinics of North America, 20,* 133–147.

Fadal, R. G. (1987). The medical management of rhinitis. In G. M. English (Ed.), *Otolaryngology: Vol. 2* (pp. 1–25). Philadelphia: Lippincott.

Forman, R., & Settle, R. G. (1990a). Burning mouth symptoms: A clinical review, part I. *Compendium of Continuing Education in Dentistry, 11,* 74–82.

Forman, R., & Settle, R. G. (1990b). Burning mouth symptoms, Part II: A clinical review. *Compendium of Continuing Education in Dentistry, 11,* 140–146.

Fox, P. C., Atkinson, J. C., Macynski, A. A., Wolff, A., Kung, D. S., Valdez, I. H., Jackson, W., Delapenha, R. A., Shiroky, J., & Baum, B. J. (1991). Pilocarpine treatment of salivary gland hypofunction and dry mouth (xerostomia). *Archives of Internal Medicine, 151,* 1149–1152.

Frank, M. E., & Smith, D. V. (1991). Electrogustometry: A simple way to test taste. In T. V. Getchell,

L. M. Bartoshuk, R. L. Doty, & J. B. Snow, Jr. (Eds.), *Smell and taste in health and disease* (pp. 503–514). New York: Raven.

Gent, J. F., Frank, M. E., & Mott, A. E. (1997). Taste testing in clinical practice. In A. M. Seiden (Ed.), *Taste and smell disorders* (pp. 146–158). New York: Thieme.

Gent, J. F., Goodspeed, R. B., Zagraniski, R. T., & Catalanotto, F. A. (1987). Taste and smell problems: Validation of questions for the clinical history. *The Yale Journal of Biology and Medicine, 60,* 27–35.

Getchell, T. V., Doty, R. L., Bartoshuk, L. M., & Snow, J. B., Jr. (Eds.) (1991). *Smell and taste in health and disease.* New York: Raven.

Gilbert, A. N., & Wysocki, C. J. (1987). The smell survey results. *National Geographic Magazine, 172,* 514–525.

Goodspeed, R. B., Gent, J. F., & Catalanotto, F. A. (1987). Chemosensory dysfunction: Clinical evaluation results from a taste and smell clinic. *Postgraduate Medicine, 81,* 251–260.

Grant, R., Miller, S., Simpson, D., Lamey, P. J., & Bone, I. (1989). The effect of chorda tympani section on ipsilateral and contralateral salivary secretion and taste in man. *Journal of Neurology, Neurosurgery, and Psychiatry, 52,* 1058–1062.

Green, B. G., & Gelhard, B. (1989). Salt as an oral irritant. *Chemical Senses, 14,* 259–271.

Henkin, R. I. (1987). Taste and smell disorders. In G. Adelman (Ed.), *Encyclopedia of neuroscience* (pp. 1185–1187). Boston: Birkhaeuser.

Henkin, R. I., Larson, A. L., & Powell, R. D. (1975). Hypogeusia, dysgeusia, hyposmia, and dysosmia following influenza-like infection. *Annals of Otology, Rhinology and Laryngology, 84,* 672–682.

Henkin, R. I., Schechter, P. J., Friedewald, W. T., Demets, D. L., & Raff, M. (1976). A double blind study of the effects of zinc sulfate on taste and smell dysfunction. *American Journal of Medical Science, 272,* 285–299.

Jafek, B. W., Eller, P. M., Esses, B. A., & Moran, D. T. (1989). Post-traumatic anosmia: Ultrastructural correlates. *Archives of Neurology, 46,* 300–304.

Jafek, B. W., Gordon, A. S. D., Moran, D. T., & Eller, P. M. (1990). Congenital anosmia. *Ear, Nose and Throat Journal, 69,* 331–337.

Jafek, B. W., Hartman, D., Eller, P. M., Johnson, E. W., Strahan, R. C., & Moran, D. T. (1990). Postviral olfactory dysfunction. *American Journal of Rhinology, 4,* 91–100.

Jafek, B. W., Moran, D. T., Eller, P. M., Rowley III, J. C., & Jafek, T. B. (1987). Steroid-dependent anosmia. *Archives of Otolaryngology, Head and Neck Surgery, 113,* 547–549.

Johnson, J. T., Gerretti, G. A., Nethery, W. J., Valdez, I. H., Fox, P. C., Ng, D., Muscoplat, C. C., & Gallagher, S. C. (1993). Oral pilocarpine for post-irradiation xerostomia in patients with head and neck cancer. *New England Journal of Medicine, 329,* 390–395.

Kalmus, H. (1971). The genetics of taste. In L. M. Beidler (Ed.), *Handbook of sensory physiology: Vol. IV, Chemical senses, Part 2: Taste* (pp. 165–179). New York: Springer-Verlag.

Keenan, C. M., Kelly, D. P., & Bogdanffy, M. S. (1990). Degeneration and recovery of rat olfactory epithelium following inhalation of dibasic esters. *Fundamental and Applied Toxicology, 15,* 381–393.

Klingmüller, D., Dewes, W., Krahe, T., Brecht, G., & Schweikert, H. (1987). Magnetic resonance imaging of the brain in patients with anosmia and hypothalamic hypogonadism (Kallmann's Syndrome). *Journal of Clinical Endocrinology and Metabolism, 65,* 581–584.

Knorr, J. R., Ragland, R. L., Brown, R. S., & Gelber, N. (1993). Kallmann syndrome: MR findings. *American Journal of Neuroscience Research, 14,* 845–851.

Kobal, G. (1982). A new method for determination of the olfactory and the trigeminal nerve's dysfunction: Olfactory (OEP) and chemical somatosensory (CSEP) evoked potentials. In A. Rothenberger (Ed.), *Event-related potentials in children* (pp. 455–461). Amsterdam: Elsevier.

Kveton, J. F., & Bartoshuk, L. M. (1994). The effect of unilateral chorda tympani damage on taste. *Laryngoscope, 104,* 25–29.

Leigh, A. D. (1943). Defects of smell after head injury. *Lancet, 244,* 38–40.

Leopold, D. A., Hornung, D. E., & Youngentob, S. L. (1991). Olfactory loss after upper respiratory infection. In T. V. Getchell, R. L. Doty, L. M. Bartoshuk, & J. B. Snow, Jr. (Eds.), *Smell and taste in health and disease* (pp. 731–734). New York: Raven.

Mackenzie, M. (1884). *A manual of diseases of the throat and nose: Vol. II: Diseases of the oesophagus, nose, and naso-pharynx.* New York: Wood.

Mahajan, S. K., Prasad, A. S., Lambujon, J., Abbasi, A. A., Briggs, W. A., & McDonald, F. D. (1980). Improvement of uremic hypogeusia by zinc: A double-blind study. *American Journal of Clinical Nutrition, 33,* 1517–1521.

Mattes, R. D., & Cowart, B. J. (1994). Dietary assessment of patients with chemosensory disorders. *Journal of the American Dietetic Association, 94,* 50–56.

Mattes, R. D., Cowart, B. J., Schiavo, M. A., Arnold, C., Garrison, B., Kare, M. R., & Lowry, L. D. (1990). Dietary evaluation of patients with smell and/or taste disorders. *American Journal of Clinical Nutrition, 51,* 233–240.

Mattes-Kulig, D. A., & Henkin, R. I. (1985). Energy and nutrient consumption of patients with dysgeusia. *Journal of the American Dietetic Association, 85,* 822–826.

Min, Y.-G., Rhee, C.-S., Choo, M.-J., Song, H.-K., & Hong, S.-C. (1994). Histopathologic changes in the olfactory epithelium in mice after exposure to sulfur dioxide. *Acta Otolaryngologica (Stockh), 114,* 447–452.

Moon, C. N., & Pullen, E. W. (1963). Effects of chorda tympani section during middle ear surgery. *Laryngoscope, 73,* 392–405.

Mott, A. E. (1991). Topical corticosteroid therapy for nasal polyposis. In T. V. Getchell, R. L. Doty, L. M. Bartoshuk, & J. B. Snow, Jr. (Eds.), *Smell and taste in health and disease* (pp. 553–572). New York: Raven.

Mott, A. E., & Leopold, D. A. (1991). Disorders in taste and smell. *Medical Clinics of North America, 75,* 1321–1353.

Murphy, C. (1985). Cognitive and chemosensory influences on age-related changes in the ability to identify blended foods. *Journal of Gerontology, 40,* 47–52.

Murphy, C., & Gilmore, M. M. (1989). Quality-specific effects of aging on the human taste system. *Perception & Psychophysics, 45,* 121–128.

Nikula, K. J., & Lewis, J. L. (1994). Olfactory mucosal lesions in F344 rats following inhalation exposure to pyridine at threshold limit value concentrations. *Fundamental and Applied Toxicology, 23,* 510–517.

Nordin, S., Monsch, A. U., & Murphy, C. (1995). Unawareness of smell loss in normal aging and Alzheimer's disease: Discrepancy between self-reported and diagnosed smell sensitivity. *Journal of Gerontology: Psychological Sciences, 50B,* P187–P192.

Osaki, T., Ohshima, M., Tomita, Y., Matsugi, N., & Nomura, Y. (1996). Clinical and physiological investigations in patients with taste abnormality. *Journal of Oral Pathology & Medicine, 25,* 38–43.

Rawson, N. E., Brand, J. G., Cowart, B. J., Lowry, L. D., Pribitkin, E. A., Rao, V. M., & Restrepo, D. (1995). Functionally mature olfactory neurons from two anosmic patients with Kallmann syndrome. *Brain Research, 681,* 58–64.

Rhodus, N. L., & Schuh, M. J. (1991). Effects of pilocarpine on salivary flow in patients with Sjögren's syndrome. *Oral Surgery, Oral Medicine, Oral Pathology, 72,* 545–549.

Rollin, H. (1978). Drug-related gustatory disorders. *Annals of Otology, 87,* 37–42.

Roydhouse, N. (1988). Retinoid therapy and anosmia. *New Zealand Medical Journal, 101,* 465.

Rugarli, E. I., & Ballabio, A. (1993). Kallman syndrome, from genetics to neurobiology. *Journal of the American Medical Association, 270,* 2713–2716.

Schechter, P. J., Friedewald, W. T., Bronzert, D. A., Raff, M. S., & Henkin, R. I. (1972). Idiopathic hypogeusia: A description of the syndrome and a single-blind study with zinc sulfate. *International Review of Neurobiology, 1(Suppl.),* 125–140.

Schiffman, S. S. (1977). Food recognition by the elderly. *Journal of Gerontology, 32,* 586–592.

Schiffman, S. S. (1979). Changes in taste and smell with age: Psychophysical aspects. In J. M. Ordy & K. Brizzee (Eds.), *Sensory systems and communication in the elderly (Aging, Vol. 10)* (pp. 227–246). New York: Raven.

Schiffman, S. S. (1983a). Taste and smell in disease (part I). *New England Journal of Medicine, 308,* 1275–1279.

Schiffman, S. S. (1983b). Taste and smell in disease (part II). *New England Journal of Medicine, 308,* 1337–1343.

Schiffman, S. S. (1991). Drugs influencing taste and smell perception. In T. V. Getchell, R. L. Doty, L. M. Bartoshuk, & J. B. Snow, Jr. (Eds.), *Smell and taste in health and disease* (pp. 845–850). New York: Raven.

Schiffman, S. S., Gatlin, L. A., Frey, A. E., Heiman, S. A., Stagner, W. C., & Cooper, D. C. (1994). Taste perception of bitter compounds in young and elderly persons: Relation to lipophilicity of bitter compounds. *Neurobiology of Aging, 15,* 743–750.

Schwartz, B. S., Doty, R. L., Monroe, C., Frye, R., & Barker, S. (1989). Olfactory function in chemical workers exposed to acrylate and methacrylate vapors. *American Journal of Public Health, 79,* 613–618.

Schwob, J. E., Leopold, D. A., Szumowski, K. E. M., & Emko, P. (1993). Histopathology of olfactory mucosa in Kallmann's syndrome. *Annals of Otology, Rhinology and Laryngology, 102,* 117–122.

Scott, A. E. (1989). Clinical characteristics of taste and smell disorders. *Ear, Nose and Throat Journal, 68,* 297–315.

Seebolm, P. M. (1978). Allergic and nonallergic rhinitis. In E. Middleton, C. Reed, & E. Ellis (Eds.), *Allergy principles and practice: Vol. 2* (pp. 868–876). St. Louis, MO: Mosby.

Sherman, A. H., Amoore, J. E., & Weigel, V. (1979). The pyridine scale for clinical measurement of olfactory threshold: A quantitative reevaluation. *Otolaryngology, Head and Neck Surgery, 87,* 717–733.

Ship, J. A., & Weiffenbach, J. M. (1993). Age, gender, medical treatment, and medication effects on smell identification. *Journal of Gerontology: Medical Sciences, 48,* M26–M32.

Singh, N., Grewal, M. S., & Austin, J. H. (1970). Familial anosmia. *Archives of Neurology, 22,* 40–44.

Smith, D. V. (1988). Assessment of patients with taste and smell disorders. *Acta Otolaryngologica, 458(Suppl.),* 129–133.

Smith, D. V. (1991). Taste and smell dysfunction. In M. M. Paparella, D. A. Shumrick, J. L. Gluckman, & W. L. Meyerhoff (Eds.), *Otolaryngology: Vol. III* (3rd ed., pp. 1911–1934). New York: Saunders.

Snow, J. B., Jr. (1983). Clinical problems in chemosensory disturbances. *American Journal of Otolaryngology, 4,* 224–227.

Stevens, J. C. (1996). Detection of tastes in mixture with other tastes: Issues of masking and aging. *Chemical Senses, 21,* 211–221.

Stevens, J. C., Bartoshuk, L. M., & Cain, W. S. (1984). Chemical senses and aging: Taste versus smell. *Chemical Senses, 9,* 167–179.

Stevens, J. C., & Cain, W. S. (1985). Age-related deficiency in the perceived strength of six odorants. *Chemical Senses, 10,* 517–529.

Stevens, J. C., & Cain, W. S. (1986). Smelling via the mouth: Effect of aging. *Perception & Psychophysics, 40,* 142–146.

Stevens, J. C., Cain, W. S., & Weinstein, D. E. (1987). Aging impairs the ability to detect gas odor. *Fire Technology, 23,* 198–204.

Stevens, J. C., & Dadarwala, A. D. (1993). Variability of olfactory threshold and its role in assessment of aging. *Perception & Psychophysics, 54,* 296–302.

Sumner, D. (1964). Post-traumatic anosmia. *Brain, 87,* 107–120.

Sumner, D. (1967). Post-traumatic ageusia. *Brain, 90,* 187–202.

Tomita, H. (1990). Zinc in taste and smell disorders. In H. Tomita (Ed.), *Trace elements in clinical medicine* (pp. 15–37). Tokyo: Springer-Verlag.

Truwit, C. L., Barkovich, A. J., Grumbach, M. M., & Martini, J. J. (1993). MR imaging of Kallmann syndrome, a genetic disorder of neuronal migration affecting the olfactory and genital systems. *American Journal of Neuroscience Research, 14,* 827–854.

Weiffenbach, J. M., Baum, B. J., & Burghauser, R. (1982). Taste thresholds: Quality specific variation with human aging. *Journal of Gerontology, 37,* 372–377.

Weiffenbach, J. M., Fox, P. C., & Baum, B. J. (1986). Taste and salivary function. *Proceedings of the National Academy of Sciences, USA, 83,* 6103–6106.

Weiffenbach, J. M., Schwartz, L. K., Atkinson, J. C., & Fox, P. C. (1995). Taste performance in Sjögren's syndrome. *Physiology & Behavior, 57,* 89–96.

Weisman, K., Christensen, E., & Dreyer, V. (1979). Zinc supplementation in alcoholic cirrhosis: A double-blind clinical trial. *Acta Medica Scandinavica, 205,* 361–366.

Wright, H. N. (1987). Characterization of olfactory dysfunction. *Archives of Otolaryngology, Head and Neck Surgery, 113,* 163–168.

Wysocki, C. J., & Gilbert, A. N. (1989). National Geographic smell survey: Effects of age are heterogenous. *Annals of the New York Academy of Sciences, 561,* 12–28.

Yamagishi, M., Hasegawa, S., & Nakano, Y. (1988). Examination and classification of human olfactory mucosa in patients with clinical olfactory disturbances. *Archives of Otorhinolaryngology, 245,* 316–320.

Yanagisawa, K., Bartoshuk, L. M., Karrer, T. A., Kveton, J. F., Catalanotto, F. A., Lehman, C. D., & Weiffenbach, J. M. (1992). Anesthesia of the chorda tympani nerve: Insights into a source of dysgeusia. [Abstract]. *Chemical Senses, 17,* 724.

Yousem, D. M., Turner, W. J. D., Li, C., Snyder, P. J., & Doty, R. L. (1993). Kallmann syndrome: MR evaluation of the olfactory system. *American Journal of Neuroscience Research, 14,* 839–843.

Zilstorff, K., & Herbild, O. (1979). Parosmia. *Acta Otolaryngologica, 360(Suppl.),* 40–41.

Zuniga, J. R., Chen, N., & Miller, I. J., Jr. (1994). Effects of chorda-lingual nerve injury and repair on human taste. *Chemical Senses, 19,* 657–665.

CHAPTER 6

The Ontogeny of
Human Flavor Perception

Julie A. Mennella
Gary K. Beauchamp

I. INTRODUCTION

The idea of childhood did not exist in medieval society. This does not mean to imply that children were neglected or despised, but rather society lacked an apparent awareness of that which distinguishes the child from the adult (Aries, 1962). Medieval art portrayed children as miniature adults, and during the middle ages and for a long time thereafter in the lower social classes, children worked and played with old and young alike (Aries, 1962; Postman, 1982). This changed greatly during the past few centuries. In part, the concern about education implied that children were not ready for life with adults. Rather, they had to be subjected to a special treatment, a sort of quarantine, before being allowed to join the adult world.

The scientific interest in the behavior of the child, especially the infant, draws into force from a number of sources, some of which date back to the turn of the twentieth century (Kagan, 1971). Infant mortality rates declined and, as a consequence, society became concerned about the psychological development of the child, not just simply his or her survival. Moreover, many believed that developmental research would provide a rational foundation on which the tenets of education, in its broadest sense, would be based.

This chapter focuses on one aspect of developmental research, that which relates to the chemical senses, with particular emphasis on food acceptance, and its implications for "education." As will be discussed, this research clearly shows that the human infant, like the adult, is not a passive receptacle for food. Rather, she or he makes active choices in accepting or rejecting certain flavors. However, these studies also reveal that the sensory world of the young infant may be quite different from that of the adult. Moreover, although the research is still far from definitive, it is possible that experience during infancy may serve to alter later responsiveness to flavors, and, in a sense, educate and civilize the young child to appreciate the flavors typical of the culture into which she or he was born.

II. SENSORY SYSTEMS THAT DETECT FLAVOR

The chemical senses, that is, the senses of taste, smell, and chemical irritation, together convey information to the infant about the overall flavor of a food. These senses not only function in the human neonate, but they change during development, probably as a function of both physical maturation and response to the environment. Because little experimental work has been done on infants' perception of chemical irritation, the following review will focus on their senses of taste and smell.

A. Taste

The sensation of taste, or gustation, occurs when chemicals stimulate taste receptors on the tongue and other parts of the oropharynx (see Figure 1). These receptors, localized in the taste buds, are innervated by portions of the facial (VIIth), glossopharyngeal (IXth), and the vagal (Xth) cranial nerves. Taste stimuli are often sepa-

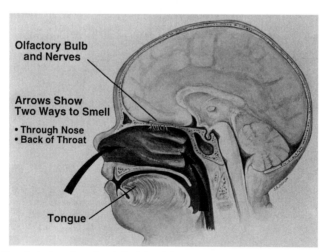

FIGURE 1   Sagittal section of an infant's head demonstrating retronasal and orthonasal routes of olfaction. [Reprinted from Mennella & Beauchamp (1993). Courtesy of *Pediatric Basics*, Vol. 65 © 1993 Gerber Products Company.]

rated into a small number of "primary" tastes: sweet, salty, bitter, sour, and perhaps savory—the taste of "umami" or monosodium glutamate (MSG).

B. Olfaction

The sensation of smell, or olfaction, occurs when chemicals stimulate the olfactory receptors in the nasal cavity. These receptors are located on the cilia of the neurons of cranial nerve 1. Unlike the sense of taste, there may be many different classes of odor stimuli, perhaps hundreds or thousands. A family of genes that may code for a large number of receptor proteins that detect odorous compounds has been identified (Buck & Axell, 1991). Sensory information from the olfactory receptor neurons is projected to the main olfactory bulbs, from there to the amygdala, and then to selected diencephalic areas, many of which are involved in modulating feeding behaviors.

Odors can reach the olfactory receptors two ways: they can either enter the nares during inhalation (orthonasal route) or travel from the back of the nasopharynx toward the roof of the nasal cavity (retronasal route) during suckling in infants and during chewing and swallowing in older children and adults (see Figure 1). Retronasal olfaction contributes significantly to the perception of flavor (Rozin, 1982). For example, holding one's nose while eating interrupts retronasal olfaction, and thereby eliminates many of the subtleties of food, leaving only the taste components (sweet, salty, sour, bitter, and savory) remaining. Another example of the importance of olfaction in flavor perception is the loss of the ability to discriminate common foods when olfactory receptors are blocked by a head cold.

III. RESPONSIVENESS OF THE FETUS AND PREMATURE INFANT TO FLAVORS

A. Taste

Taste cells first appear in the human fetus early in gestation, and morphologically mature cells appear at about 14 weeks (Bradley, 1972). Chemical present in the amniotic fluid may stimulate the fetal taste receptors when the fetus begins to swallow episodically at about the 12th week of gestation (Conel, 1939; Liley, 1972; Pritchard, 1965). The chemical composition of the amniotic fluid varies over the course of gestation, particularly as the fetus begins to urinate. By term, the human fetus has swallowed significant amounts of amniotic fluid (200–760 ml daily) and has been exposed to a variety of substances, including glucose, fructose, lactic acid, pyruvic acid, citric acid, fatty acids, phospholipids, creatinine, urea, uric acid, amino acids, proteins, and salts (Liley, 1972). Differential fetal swallowing following the injection of sweet or bitter substances into the amniotic fluid (Liley, 1972, DeSnoo, 1937) suggests that the fetus shows a preference for sweet and a rejection of bitter;

these observations are inconclusive because of the methodological limitations in measuring fetal responses, however.

Studies on taste sensation in the preterm infant suggest that sweet taste preference is evidenced prior to birth. When preterm infants, who had been fed exclusively via gastric tube, were presented intraorally with small amounts of either glucose solutions or water, they exhibited more nonnutritive sucking in response to the glucose compared to the water (Tatzer, Schubert, Timischl, & Simbruner, 1985). Because premature infants are at risk for aspirating fluids due to an immature suck–swallow coordination, a methodology was developed that did not necessitate the delivery of any fluids while administering a taste (Maone, Mattes, Bernbaum, & Beauchamp, 1990). Here the taste substance was embedded in a nipple-shaped gelatin medium that released small amounts of the substance when it is mouthed or sucked. Infants born preterm and tested between 33 and 40 weeks postconception produced more frequent, stronger sucking responses when offered a sucrose-sweetened nipple compared with a latex nipple.

B. Olfaction

Although the olfactory system is well developed prior to birth (Arey, 1930; Bossey, 1980; Nakashima, Kimmelman, & Snow, 1985), it is not known whether the fetus responds to olfactory stimuli. However, reports indicate that the environment in which the fetus lives—the amniotic fluid—can indeed be odorous. Not only does its odor indicate certain disease states (Mace, Goodman, Centerwall, & Chinnock, 1976), but it also reflects the types of foods eaten by the pregnant mother (Hauser, Chitayat, Berbs, Braver, & Mulbauer, 1985). In another study, amniotic fluid samples were obtained from 10 pregnant women who were undergoing routine amniocentesis procedure (Mennella, Johnson, & Beauchamp, 1995). Approximately 45 min prior to the procedure, five of the women ingested placebo capsules, whereas the remaining five ingested capsules containing the essential oil of garlic. Randomly selected pairs of samples, one from a woman who ingested garlic and the other from a woman who ingested placebo capsules, were then evaluated by a sensory panel of adults. The odor of the amniotic fluid obtained from four of the five women who had ingested the garlic capsules was judged to be stronger or more like garlic than were the paired samples collected from the women consuming placebo capsules (see Figure 2). Thus garlic ingestion by pregnant women significantly altered the odor of their amniotic fluid.

Because the normal fetus swallows significant amounts of amniotic fluid during the latter stages of gestation (Pritchard, 1965) and has open airway passages that are bathed in amniotic fluid (Schaffer, 1910), the fetus may be exposed to a unique olfactory environment prior to birth. Studies on other animals reveal that certain odors that were experienced *in utero* were preferred postnatally (e.g., Hepper, 1987); whether similar mechanisms operate in humans remain unknown.

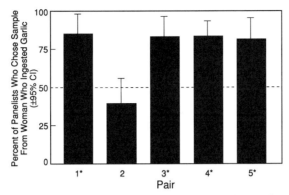

FIGURE 2   Percentage of panelists who chose an amniotic fluid sample as smelling stronger or more like garlic than another sample. According to a forced-choice paradigm, panelists were presented individually with paired samples of amniotic fluid; one of the pair was obtained from a woman who had ingested garlic capsules and the other was from a woman who ingested placebo capsules. Panelists were asked to indicate which of the pair smelled "more like garlic" or "stronger." A value of 50% would be expected if there were no difference in the odor of the samples and hence the panelists responded at random. (Data represent mean $\pm$ 95% CI; *, $p < .02$.)

IV. TASTE AND OLFACTORY PERCEPTION DURING HUMAN INFANCY

A. Taste

In some of the earliest investigations on human taste development, facial expressions, suggestive of contentment and liking or discomfort and rejection, were used to assess the responsiveness of newborns to taste stimuli. During the first few hours of life, infants display relatively consistent, quality-specific facial expressions when the sweet taste of sugars (facial relaxation, sucking movements), the sour taste of concentrated citric acid (facial grimaces), and the bitter taste of concentrated quinine and urea (tongue protrusions, facial grimaces) are present into the oral cavity (Steiner, 1977; Rosenstein & Oster, 1990; Ganchrow, Steiner, & Munif, 1983). No distinct facial response is evidenced for salt taste (Steiner, 1977; Rosenstein & Oster, 1990). Infants also display distinct positive facial expressions, similar to that observed with sweetness, when tasting MSG-flavored soup as compared to their response when tasting the soup diluent alone (Steiner, 1987). That MSG alone does not appear to elicit positive facial responses raises the question of exactly what it is about the MSG-flavored soup that is of apparent positive hedonic valence.

The most frequently used method to study taste preferences involves comparing how much the infant consumes of a taste solution and the diluent solution during brief presentations; most of these intake studies used weaker concentrations of taste stimuli than did the studies on facial expressions previously mentioned. If the infant ingests more of the taste solution than the diluent, one can infer that (1) the infant can detect the taste and, with less certainty, (2) the infant prefers or likes the tastant more than the diluent. Generally, the diluent is water; however, slightly sweetened water has been used as the diluent for testing the infant's response to sour and bitter tastes. Consistent with the finding for premature infants, newborn infants exhibit a strong acceptance of sweet-tasting sugars. They will respond even to dilute sweet solutions and can differentiate varying degrees of sweetness (Desor, Maller, & Turner, 1977). In contrast, newborns reject the sour taste of citric acid, whereas they are apparently indifferent to the taste of low to moderate concentrations of salt and the bitter taste of urea (Desor, Maller, & Andrews, 1975).

Finally, as was the case for premature infants, the study of the newborn infant's sucking patterns has been utilized as an alternative to the measurement for intake. Such studies also indicate that sweet stimuli enhance suckling and thus are positive and salt taste is negative in that it suppresses sucking relative to the infant's response to the diluent alone (Crook, 1978).

Although each measure has its limitations, the convergence of research findings supports the conclusion that newborns are quite sensitive to and prefer sweet taste, whereas they reject substances having a strong sour taste. Conclusions regarding the neonate's responses to bitter and salty tastes are more problematic, however. Newborns respond with highly negative facial expressions to concentrated quinine and urea, but they do not reject moderate concentrations of urea (Desor et al., 1975). The reasons for this differential response remain unclear. Perhaps the newborn can detect bitter substances, but the ability to reject a substance or modulate intake requires further maturation. Further studies using a variety of bitter stimuli and additional behavioral measures could resolve this question. Additionally, bitter-tasting chemicals may interact with one or more of several different transduction pathways (Spielman, Huque, Nagai, Whitney, & Brand, 1994). The newborn infant may be differentially sensitive to different bitter compounds due to different rates of maturation of these transductive sequences.

With regard to salt perception, studies measuring intake and facial expressions suggest that the newborn infant is indifferent to and may not detect salt. However, salt does appear to suppress some parameters of sucking in newborns. Here too, more research is needed to clarify the newborn's response to salt. However, no studies suggest that the taste of salt is attractive to the newborn infant.

Babies beyond the neonatal period (1 to 24 months) have been most neglected in studies on taste development. Nonetheless, a few findings suggest that changes in taste responses occur during this time in development. Studies conducted in Mexico focused on the responses of well-nourished and malnourished infants, aged 2 to 24 months, to determine whether the protein-calorie status of the infants affected their taste preferences (Vasquez, Pearson, & Beauchamp, 1982). At all ages, both the well-nourished and malnourished infants preferred the sucrose but rejected the bitter (urea) and sour (citric acid) stimuli. The infants under the age of 1 year preferred the salty solutions whereas those older than one year were indifferent to the salt.

The salt preference of infants is of considerable interest. Recall that newborn infants were indifferent to or rejected salt relative to plain water. The suggestion of a developmental shift in salt acceptability has been supported in more recent studies of children from the United States, where preferential ingestion of salt water relative to plain water first emerged at approximately 4 months of age (Beauchamp, Cowart, & Moran, 1986). Additionally, studies that measured the infants' patterning of suckling as well as intake demonstrated age-related changes in the response to salt that were consistent with an increased sensitivity occurring at 3 to 6 months of age (Beauchamp, Cowart, Mennella, & Marsh, 1994; Harris & Booth, 1987). It has been argued that experience with salty tastes probably does not play a major role in the shift from indifference or rejection of salt at birth to acceptance in later infancy (Beauchamp & Cowart, 1990). Rather, this change in response may reflect postnatal maturation of central and/or peripheral mechanisms underlying salt taste perception. Thus, the salt preference that emerges at approximately 4 months of age may be largely unlearned. Although animal model studies demonstrate that early alterations in sodium balance alter long-term salt preference behavior (Hill & Mistretta, 1990), whether this occurs in humans is unknown but remains an important public health issue that needs to be addressed.

Well-nourished and malnourished infants were also tested with a mixture of amino acids and peptides (casein hydrolysate) and with MSG (Vasquez et al., 1982). For these two taste substances, a low-glutamate soup was used as the diluent. Both malnourished and well-nourished infants, as young as 2 to 4 months of age, preferred the MSG-flavored soup when compared to the soup alone, and there was no difference between the groups. In contrast, the nutritional status of the infants influenced their responses to the casein hydrolysate. For the malnourished infants, soup with casein hydrolysate was preferred to soup alone; the opposite response was observed for well-nourished infants and for recovered malnourished infants. Thus, chemosensory response to flavors may be influenced by the nutritional needs of the infant.

Finally, a study on the sensitivity of infants to bitter taste revealed that relatively low concentrations of urea were not rejected in newborn infants but rejection was evident among infants who were 14–180 days of age (Kajuira, Cowart, & Beauchamp, 1992). These data are consistent with the idea that there is an early developmental change in bitter perception or the ability to regulate the intake of bitter solutions.

In summary, the sensory world of the young infant is different from that of the adult; sensitivity to several different taste stimuli appears to develop at different times postnatally (Table 1). Specifically, responses to sweet tastes are evident prenatally and major changes are not known to occur postnatally. The rejection of sour taste is evidenced from birth onward. In contrast, salt and bitter sensitivities change post-

TABLE 1   Developmental Changes in Responses of Infants to Tastes[a]

Taste	Premature infants	Newborns	Infants (1–24 months)
Sweet sugars	Preference	Preference	Preference
Sour			
citric acid	Not known	Rejection	Rejection
Bitter			
Quinine	Not known	Rejection	Not known
Urea	Not known	Indifference, rejection[b]	Rejection
Salty (NaCl)	Not known	Indifference, rejection[c]	Indifference, rejection, preference[e]
Savory (MSG)	Not known	Preference[d]	Preference[d]

[a] Infants' responses to various taste solutions relative to water or diluent.

[b] Facial expressions suggest rejection, whereas intake studies suggest indifference.

[c] Sucking measures suggest rejection, whereas intake and facial expression studies suggest indifference.

[d] Preference seen only when monosodium glutamate (MSG) mixed with soup; MSG solution alone is rejected relative to plain water.

[e] Preference emerges at approximately 4 months of age; prior to that, indifference or rejection depending on the methods used.

natally, with salt taste providing the clearest example. The developmental changes in salt preferences may reflect changes in sensitivity independent of experience, or they may be a consequence of specific experiences, or both.

B. Olfaction

Shortly after birth, human infants can detect a wide variety of volatile chemicals and can discriminate among them, and are capable of complex associative learning with odors. We have purposely not discussed in detail the research on the role of smell in mother–young attachments and its possible role in later social and sexual behaviors because this material has been reviewed elsewhere (Leon, 1992; Schaal, 1988; Wilson & Sullivan, 1994). A key question in the area of human olfaction is whether hedonic responses to any odors are present at birth or whether all are acquired postnatally, presumably as a function of experience (Engen, 1982). Two fundamental problems exist in addressing this question, however. First, how does one measure affective reactions to odors in the preverbal child? And second, what odors should be used?

Some investigators have measured movements toward or away from an odor source and have concluded that such responses indicated olfactory preferences and aversions, respectively. If this interpretation is warranted, then responses to odors are evidenced shortly after birth (e.g., Engen, Lipsitt, & Kaye, 1963; Peterson & Rainey, 1910–1911; Reiser, Yonas, & Wilkner, 1976; Self, Horowits, & Paden, 1972; Sarnat, 1978). However, it is not certain whether the infants' approach or avoidance response was the result of olfactory or trigeminal stimulation, because few of the odors used in these studies were purely olfactory stimulated (Doty, 1986).

Perhaps the most salient of odors for the newborn infant are those originating from the mother (MacFarlane, 1975). Within hours after birth, mothers and infants can recognize each other through the sense of smell alone (see Schaal, 1988, for review). Breast-fed infants spent more time orienting toward a breast pad previously worn by their lactating mothers than one worn by an unfamiliar lactating woman (Schaal, 1986, 1988). They move their head and arms less and suck more when they are exposed to their mothers' odors (Schaal, 1986). This ability of breast-fed infants to discriminate the odors of their mothers from those of other lactating women is not limited to odors emanating from the breast region, because they can also discriminate odors originating from their mothers' underarms (Cernoch & Porter, 1985) and neck (Schaal, 1986). A recent study revealed that newborns prefer their mother's breast when it is unwashed as compared to when it had been thoroughly washed and thereby devoid of some of its natural odors (Varendi, Porter, & Winberg, 1994). Thus, like other mammalian young, maternal odors may play a role in guiding the infant to the nipple area and in facilitating early nipple attachment and suckling.

V. EARLY RESPONSIVENESS BY THE HUMAN INFANT TO FLAVOR: EXAMPLE OF MOTHER'S MILK

A. Folklore on the Choice of Wet Nurse and Infant Feeding

Before sanitary precautions were known, and before the differences between human milk and the raw milk of other animals were understood, the only successful substitute for a mother's milk was the milk of another woman. In fact, many women gave up the care of their own children to nurse those of their social superiors because the occupation of wet nursing was quite lucrative before its decline in the 18th century (Fildes, 1986). But as wet nursing flourished, so too did superstitions. Wet nurses were selected carefully because of the ancient belief that the child could imbibe the characteristics of the woman's personality through her milk (Soranus, 1st/ 2nd century A.D.). If the woman hired was promiscuous or evil-tempered, the child would suck in these vices through the milk (Mixsell, 1916). And a child who suckled from a drunken wet nurse would surely grow up to be the same—witness the case of the bibulous Emperor Nero (Wickes, 1953).

Not only was the woman's character of concern but so too were the sensory attributes of her milk. The milk, if it was good, should be white, have a good smell and sweet taste, and be of moderate consistency. Nursing women were advised to eat certain foods and avoid others because of the belief that the woman's diet could affect the composition and flavor of her milk. For example, it was said that an offensive odor in the woman's milk could be made more pleasant by having the woman drink fragrant wines or eat sweet foods (Ebers Papyrus, circa 1550 B.C.). Because the woman's milk was believed to be converted blood, the physical characteristics that were sought in the wet nurse, and the diet and regimen prescribed once she was hired, were intended to ensure that she had a healthy supply of blood. Moreover, mother's milk was believed to be best suited for the child because it was the same blood that provided nourishment in the womb and with which the child was familiar (Ettmueller, 1699; Guillemeau, 1635; Fildes, 1986). Clearly, that a woman's diet could influence the composition and flavor of her milk and that substances in human milk could be transmitted from the wet nurse or mother and have long-lasting effects on the child is not a new notion. During the past few years, new research has focused on the early sensory experiences of the human infant, and the medium for these sensory experiences is mother's milk. Following a brief discussion of the related research on other animals, work on the human mother–infant dyad will be highlighted.

B. Transfer of Volatiles to Milk: Dairy Cattle

In 1757, a treatise was published that described how cows fed beets and turnips produced a bitter-flavored milk (Bradley, 1757). Whether this altered flavor was due to an actual change in the taste or odor of the milk is unknown. Since this time, there have been numerous reports that a variety of odors from various feeds (e.g., silage, rye) and weeds (e.g., garlic, onion) eaten by the cow, or from the air it breathes, can be transmitted to the milk while it is in the udder (Babcock, 1938; Bassette, Fung, & Mantha, 1986). Some of the volatile components associated with these flavors, such as dimethyl sulfide, have been identified (Shipe et al., 1978).

"Off-flavors" in milk continue to be an important issue for the dairy industry because they curtail the consumption of dairy products (Bassette et al., 1986). In 1978 the American Dairy Science Association standardized the language used in the sensory evaluation of milk and developed a classification system based on the causes of the off-flavors (Shipe et al., 1978; Shipe, 1980). In addition to transmitted flavors, off-flavors also result from heat treatment, exposure to light, lipolysis, oxidation, the accumulation of the products of bacterial metabolism, or some other yet unknown cause.

Perhaps the most common and readily recognized transmitted off-flavor is that which is produced when the cow eats and grazes on wild garlic or onions. To determine the mechanisms underlying the transmission of flavors from the cow's feed or environment, milk samples were obtained before and after onion odors were introduced directly into the cow's lungs or rumen by means of tracheal and ruminal fistulae, respectively. An evaluation of the milk samples by a trained sensory panel revealed that volatiles can be transferred to the udder via vascular routes from either the digestive or the respiratory system (Dougherty et al., 1962; Shipe et al., 1962).

Interestingly, the off-flavors produced when a cow eats onion are not described as onionlike (Dougherty et al., 1962), but are described as garliclike when the cow eats wild garlic (Shipe et al., 1962, 1978).

In summary, the research on dairy cows reveals that a wide variety of flavors can be transmitted to the udder, which, in turn, alters the milk flavor. These sensory changes in mother's milk may be a potential source of chemosensory information for the suckling animal, enabling it to learn about the dietary choices of the mother.

C. Effects of Early Flavor Experiences on the Behavior of Nurslings

Research on other mammals suggests that the young develop preferences for the flavors of the foods eaten by the mother during nursing. The growth rate of weanling pigs improves when a flavor that had been incorporated into the sow's feed during lactation was added to the weanling's feed (Campbell, 1976). Moreover, weanling animals actively seek and prefer the flavor of the diet eaten by the mother during nursing (see Table 2) (Bilkó, Altbacker, & Hudson, 1994; Capretta & Rawls, 1974; Galef & Clark, 1972; Galef & Henderson, 1972; Galef & Sherry, 1973; Gullberg, Ferrel, & Christensen, 1986; Hunt, Kraebel, Rabine, Spear, & Spear, 1993; London, Snowdon, & Smithana, 1979; Mainardi, Poli, & Valsecchi, 1989; Morrill & Dayton, 1978; Phillips & Stainbrook, 1976; Valsecchi, Moles, & Mainardi, 1993; Wuensch, 1978) and are more likely to accept unfamiliar flavors if they experience a variety of different flavors during the nursing period (Capretta, Petersik, & Stew-

TABLE 2   Summary of Experiments Suggesting Experience with Flavors in Mother's Milk Affects Postweaning Preferences

Type of Flavor	Species	References
Lab chow	*Rattus norvegicus*	Galef and Clark, 1972
		Galef and Sherry, 1973
		Galef and Henderson, 1972
Garlic	*Rattus norvegicus*	Capretta and Rawls, 1974
Chocolate, vanilla, rum, and walnut	*Rattus norvegicus*	Capretta et al., 1975
Alcohol	*Rattus norvegicus*	Phillips and Stainbrook, 1976
		Hunt et al., 1993
Onion	*Rattus norvegicus*	Wuensch, 1978
Citric acid	*Rattus norvegicus*	London et al., 1979
Caffeine	*Rattus norvegicus*	Gullberg et al., 1986
Fennel	*Mus musculus*	Maniardi et al., 1989
		Valsecchi et al., 1993
Firanor #3	*Sus scrofa*	Campbell, 1976
Butterscotch, maple, and ethyl butyrate	*Bos taurus*	Morrill and Dayton, 1978
Onion	*Ovis aries*	Nolte and Provenza, 1991
Garlic	*Ovis aries*	Nolte and Provenza, 1991
Cumin	*Ovis aries*	Schaal, Orgear, and Porter, 1994
Juniper	*Oryctolagus cuniculus*	Bilkó et al., 1994

ard, 1975). Although the milk was not evaluated before and after the lactating females consumed the flavor or flavors, the research does suggest that the young animals were learning via flavor cues in the mother's milk, rather than from other sensory cues present in the mother's breath or body odor, or from food particles clinging to the mother's fur or vibrissae (Galef & Henderson, 1972; Wuensch, 1978; Nolte & Provenza, 1991).

The majority of the studies listed in Table 2 evaluated the preference of young animals for a short period of time during weaning. Some of these preferences were short-lived (Galef & Henderson, 1972). However, stronger and more persistent preferences were evidenced when the young animal was exposed to the flavors during both the nursing and the postweaning periods (Capretta & Rawls, 1974). Whether there are sensitive periods for exposure to flavors in mother's milk remain unknown (see Provenza & Balph, 1987).

At weaning, the young animal is faced with learning what to eat and how to forage. An important factor influencing the weanling's dietary choices appears to be exposure to the diet of adult conspecifics (Galef, 1971). Although pups are attracted to the feeding site of adult conspecifics (Galef & Clark, 1971, 1972), the mother appears to be more important than other conspecifics in influencing the dietary habits of her offspring. The lambs that ate mountain mahogany with their mothers formed more persistent preferences for mountain mahogany than did lambs exposed to the shrub alone (Nolte, Provenza, & Balph, 1990). Similarly, lambs that ate novel foods with their mothers consumed approximately twice as much of these foods after weaning than did lambs that ate the foods with a dry ewe (Thorhalldottir, Provenza, & Balph, 1990).

In summary, exposure to flavors in mother's milk may be one of several ways in which the mother teaches her young what foods are "safe" (Rozin, 1976). Consequently, young animals may tend to choose a diet similar to that of the mothers when faced with their first solid meal.

D. Flavor of Human Milk and Effects on the Infant

In a study on the flavor properties of human milk (Barker, 1980; McDaniel, 1980), fore- and hindmilk samples were collected from 24 lactating women during the morning hours on three consecutive days. Within three hours of expression, a trained sensory panel evaluated the milk samples for the taste quality of sweetness and for textural properties such as viscosity and mouth-coating. Each of these sensory attributes varied from mother to mother and from foremilk to hindmilk, with the primary taste quality of the milk being its sweetness; human milk was estimated to be as sweet as a 2.1% lactose solution.

The sensory panel also reported that off-flavors were present in 30% of the samples. As discussed previously, these off-flavors could have been transmitted from the mother's diet or could have resulted from procedures such as handling. Of particular interest was the finding that panelists used the verbal descriptors *hot, spicy,* and *peppery* to describe the milk of one woman who had consumed a "spicy" meal during the test period.

1. Response of Infants to Transfer of Garlic to Mother's Milk

Some studies have systematically explored whether flavors from a woman's diet are transmitted to her milk, and what effects, if any, this has on the behavior of her breast-fed infant (Mennella & Beauchamp, 1991a). Mothers and their exclusively breast-fed infants were tested on 2 days separated by 1 week. On one testing day, the mother ingested garlic capsules whereas on the other testing day she ingested placebo capsules. Each infant fed on demand, was weighed immediately before and after each feed to determine the amount of milk consumed, and was videotaped to determine total time attached to their mother's nipples. Milk samples, obtained from each woman at fixed intervals before and after the ingestion of the capsules, were evaluated by a sensory panel of adults within hours of expression.

Like the milk of other mammals, volatiles from a woman's diet are transmitted to her milk. Maternal garlic ingestion significantly and consistently increased the perceived intensity of the milk odor; this increase in odor intensity peaked in strength 2 hr after ingestion and decreased thereafter. There was no perceived change in the odor of the milk on the days the mothers ingested the placebo capsules. The infants breast fed longer and sucked more when the milk smelled like garlic as compared to when this flavor was absent, thus suggesting that they detected the sensory change in their mother's milk. There was a tendency for the infants to ingest more milk as well.

Another study revealed that repeated consumption of garlic by nursing mothers modified their infants' response to garlic-flavored milk (Mennella & Beauchamp, 1993b). That is, the infants of mothers who had repeatedly consumed the garlic capsules breast fed for similar periods of time during the 4-hr test session in which their mothers consumed garlic as compared with the session in which their mothers ingested the placebo. In contrast, the infants who had no or minimal exposure to garlic volatiles in their mother's milk spent more time breast feeding when their mothers ingested garlic than when their mothers ingested the placebo, thus corroborating previous findings (Mennella & Beauchamp, 1991a).

Perhaps the garlic flavor became monotonous to those infants who were repeatedly exposed to it in mother's milk. Over the short term, children (Birch & Deysher, 1986) and adults (Rolls, Rowe, & Rolls, 1982) report that the palatability of a food, and the amount of it consumed, declines following repeated consumption of that food, whereas less recently consumed foods are considered more palatable and stimulate food intake. Moreover, the garlic-flavored milk may have aroused the infants who were exposed to a diet of mother's milk relatively low in flavor or garliclike compounds. When aroused, mammalian infants will suck more (Bridger, 1962) and exhibit a variety of other oral behaviors (Korner, Chuck, & Dontchos, 1968; Terry & Johanson, 1987). This enhanced suckling time did not result in an increase in milk intake, however. The absence of an effect on the infant's milk intake may reflect milk availability, infant capacity, or both. Nonetheless, these findings suggests that the sensory attributes of mother's milk contribute to the patterning and duration of suckling at the breast.

2. Response of Infants to Vanilla Flavors in Mother's Milk and Formula

Although much research will be required to understand fully the effects of exposure to flavors in mother's milk on the infant's behavior, several points are clear. First, human milk is not a food of invariant flavors (Mennella & Beauchamp, 1991a, 1991b, 1993a, 1996). Rather, it provides the potential for a rich source of varying chemosensory experiences to the infant. Second, the prior flavor experiences of mothers, and consequently their infants, may modify the infants' responses to these flavors. Thus, these findings might imply that the sensory world of the breast-fed infant is very rich, varied, and quite different from that of the bottle-fed infant. Bottle-fed infants, who experience a constant set of flavors from standard formulas, may be missing significant sensory experiences that, until recent times in human history, were common to all infants.

Recent work in our laboratory revealed that vanilla, one of the most popular flavors in the world and widely used by many cultures (Pangborn, Guinard, & Davis, 1988), is also transferred to human milk (Mennella & Beauchamp, 1994). Because some authorities have recently recommended that vanilla be added to follow-on formulas intended for infants beginning the weaning process, reportedly for both its flavoring properties and its ability to provide variety to the bottle-fed infant's otherwise "bland" diet (Food Advisory Committee Report on the Review of Additives in Foods Specially Prepared for Infants and Children, 1992) and because vanilla-flavored pacifiers (Ross Laboratories, Inc., Columbus, Ohio) are currently being distributed in some hospitals, two experiments were designed to investigate the responses of breast- and formula-fed infants to vanilla flavors in mother's milk or formula, respectively (Mennella & Beauchamp, 1996).

Consistent with what was reported previously with garlic, the nursing mother's consumption of a flavor altered the behavior of her infant during breast feeding. That is, the infants breast fed longer and consumed more milk when it was flavored with vanilla. When studying how a change in the flavor of mother's milk affects the behavior of the infant at the breast, it is difficult to separate the direct effects on the infant from other possible influences the consumed flavors could have on the mother (e.g., changes in the odor of the mother's breath or sweat) (see Mennella & Beauchamp, 1991a, for further discussion). Consequently, one cannot unequivocally conclude that the flavor change in the mother's milk was responsible for the alteration of the infant's suckling behavior. To examine this issue directly, one could study the breast-fed infant's suckling response when feeding mother's milk (unaltered or flavored) from a bottle. However, there are several methodological constraints with this approach, and many breast-fed infants do not have experience consuming mother's milk from a bottle (Auerbach & Danner, 1988).

The second experiment was designed to investigate only the effects of vanilla flavoring on the exclusively formula-fed infant's feeding behaviors by adding the flavor directly into the infant's formula. Consistent with what was found for the breast-fed infant, the response of bottle-fed infants to the flavored formula was altered relative to their response to the unflavored formula. In the first test (short-term preference), the infants sucked more vigorously when feeding the vanilla-flavored formula, and in the second test, which encompassed an entire feeding, they spent more time feeding on the first bottle when it contained the vanilla flavor. These data, along with those reported previously (Mennella & Beauchamp, 1991a, 1991b, 1993a, 1993b), strongly support the hypothesis that flavors, either consumed by the mother and transmitted to her milk or added to formula, are detected by the infant and serve to modulate feeding.

Why did the infants respond to the vanilla-flavored milk by enhanced suckling? As before (Mennella & Beauchamp, 1991a), the nursing mothers in the study were also asked to eat bland diets devoid of vanilla flavor during the 3 days preceding each testing day. Perhaps the novelty was sufficient to induce increased suckling or the flavor of vanilla may be inherently positive, either as a result of exposure or, like sweet taste, as a hard-wired (innate) response.

These data support the hypothesis that flavors, either consumed by the mother and transmitted to her milk or added to formula, are detected by the infant and serve to modulate feeding. They also suggest that experience with a flavor in milk alters the infant's responsiveness to that flavor during subsequent feedings. Whether experience with vanilla-flavored pacifiers affects the infant's response to similarly flavored mother's milk or formula needs to be examined experimentally.

3. Response of Infants to Transfer of Alcohol to Mother's Milk

The complexities of flavor transfer from a mother to her infant, and the possible effects on the infant's suckling behaviors, are exemplified in studies on maternal alcohol consumption. For centuries, lactating women have been given advice about drinking alcohol. Some advice claimed that the most frequent source of acquired alcoholism was exposure to alcohol in mother's milk, whereas other recommendations were that the mother or wet nurse should drink alcoholic beverages, especially beer, to increase milk supply and strengthen the breast-feeding infant (Robinovitch, 1903; Routh, 1879). In response to the latter folklore, beer companies marketed low-alcoholic beers, or "tonics," during the early 1900s as a means for women to stimulate their appetite, increase their strength, and enhance their milk yield (Krebs, 1953).

Using methods similar to that described for garlic and vanilla, it was found that the odor of human milk is altered when nursing women drink a small dose of alcohol in orange juice (Mennella & Beauchamp, 1991b) or in beer (Mennella & Beauchamp, 1993a). The sensory change in the milk parallels the changing concentrations of ethanol in the milk. It was also found that, contrary to the folklore, breast-fed infants consumed less milk during the 3 to 4 hours after their nursing mothers drank a small dose of alcohol (Mennella & Beauchamp, 1991b, 1993a).

Several factors could account for this decrease in milk intake. First, the infants may be responding to the change in the flavoring of their mother's milk. That is, for reasons as yet unknown, the flavor of alcohol may be relatively unpalatable. However, this seems unlikely because low levels of alcohol as evidenced in the milk (Mennella & Beauchamp, 1991b) are often described as sweet and pleasant.

A second explanation, that the depression in milk intake was due to a pharmacological effect of alcohol on the infant, is also unlikely because the infants tended to consume less milk during the first feeding that followed their mother's consumption of an alcoholic beverage (Mennella & Beauchamp, 1993a). Whether short-term exposure to small amounts of alcohol affects the infants in other ways requires further study. However, it is known that repeated exposure to small amounts of alcohol in breast milk is associated with a slight but significant detrimental effect on the infant's motor development at 1 year of age (Little, Anderson, Ervin, Worthington-Roberts, & Clarren, 1989).

A third explanation is that the decrease in the infant's milk intake may represent a pharmacological effect of alcohol on the nursing mothers. That is, ethanol consumption may affect the milk-ejection reflex, although a previous study concluded this not to be the case for the dosage used in the present study (Cobo, 1973). Moreover, ethanol consumption may alter the composition of milk such that, although infants consume less milk, they take in the same number of calories. Chronic ethanol consumption by lactating rat dams during both pregnancy and lactation resulted in milk that was higher in lipid and lower in lactose content, when compared to the milk of control rat dams (Vilaró, Viñas, Remesar, & Herrera, 1987). Because no alteration in protein or water content was observed, the milk of ethanol-exposed rats had a higher energy content due to the greater energetic value of lipids compared to proteins and lactose. Whether this altered milk composition was due to a direct consequence of ethanol intake or malnutrition is not known. Nor is it known whether ethanol intake, in the short term, has similar effects on milk composition and yield in humans, or whether, by relaxing the mother, it facilitates the let-down reflex. Future research is needed to address these issues.

VI. CONCLUSIONS AND FUTURE DIRECTIONS

This chapter focused on the ontogeny of taste and smell perception as it relates to the feeding behavior of infants and young children. We have argued here that the fetus and newborn infant have functioning chemosensory systems and that their feeding and expressive behaviors are modulated by taste and smell stimuli. Yet although these sensory systems are operable very early in ontogeny, the human fetus and newborn infant are not merely miniature adults; their sensory systems mature postnatally and are likely influenced by experiences in ways not yet fully understood.

There are a number of major gaps in our understanding of the ontogeny of smell, taste, and feeding in humans. First, there is no research on whether sensory experiences *in utero* serve to modulate postnatal responsiveness to tastes and smells in humans, although animal model studies, as reviewed earlier, provide strong suggestive evidence. Moreover, whether the apparent redundancy in exposure to flavors in amniotic fluid and mother's milk ensures that the young animal can acquire preferences for a variety of foods eaten by the mother during different stages of development (see Bilkó et al., 1994) is clearly an area in which further research is needed.

Second, and related to the first point, there is also remarkably little research on the short-term and long-term effects of early experience with the taste and smell of foods on flavor preferences and food choices. These questions not only have ramifications for our understanding of the development of food habits but are also relevant for our understanding of nutritional concerns such as excessive salt and fat consumption, which may lead to hypertension and obesity, respectively, and the effects on later preferences of early exposure to substances such as alcohol and tobacco.

A third area that needs attention is the source of the large individual differences among infants and children in response to and preference for tastes and smells. Although there are clearly consistencies across infants (e.g., preference for sweet tastes, general avoidance of bitter tastes), the differences among infants are often as striking as the similarities (c.f. Davis, 1928). What roles the genetic differences among individuals play have not been determined. However, it is known that the ability to taste some compounds, such as phenylthiocarbamide (Blakeslee, 1932; Snyder, 1931), and to smell others, such as androstenone (Wysocki & Beauchamp, 1984), has a genetic component, but how these kinds of genetic differences interact with experience in influencing food choice and flavor preference in infants and young children remains obscure.

Food habits and preferences are said to be among the last characteristic of a culture that is lost during the immigration of an individual or group into a new culture (Rozin, 1980). This would seem to imply that the kinds of foods chosen are firmly established early in life. Yet there is little evidence to support this belief (reviews: Rozin, 1984; Ray & Klesges, 1993). For example, similarities between the food preferences of children and their parents are often small (Birch, 1980; Pliner & Pelchat, 1986; Burt & Hertzler, 1978), and indeed, there is some evidence in primates and in humans that the younger individuals are more willing to sample novel foods (review: Beauchamp & Maller, 1977).

How can these seemingly contradictory tendencies be reconciled? Low levels of positive correlation between the food preferences of parents, especially the mother, and their children may not mean that early experience has no effect on later flavor choice and preferences. As noted, the sensory systems of infants, children, and adults are different, so that one might not expect similar choices even if there were experiential effects of exposure. Second, there is no standardized method of measuring what constitutes a food preference agreement (review: Ray & Klesges, 1993; Birch, 1979). Moreover, no studies, to our knowledge, have evaluated the preferences for flavors, not foods, within and among families. Family studies that evaluate both the kinds of flavors which are preferred, chosen, or avoided, and the intensity of those flavors most preferred need to be done, and such research should assess the preferences of children directly because mothers' accounts of their children's preferences are often inaccurate and biased (see Birch, 1980). Finally, experimental studies, even with animal models, have often neglected the very early exposure to flavors that occurs *in utero* and via mother's milk.

As we have suggested (Mennella & Beauchamp, 1991a, 1993b, 1996b; Mennella, 1995), early sensory experiences may be particularly important in human development, and the advent of formula feeding may not only deprive infants of important immunological and perhaps psychological benefits (review: Goldman, Atkinson, & Hanson, 1987), but may also limit their exposure to an important source of information and education about the flavor world of their mother, family, and culture.

Acknowledgments

This research was supported in part by Grants AA09523 and DC00882 from the National Institutes of Health and a grant from the Gerber Companies Foundation.

References

Aries, P. (1962). *Centuries of Childhood: a social history of family life.* [trans. from the French by Robert Baldick]. New York: Knopf.

Arey, L. B. (1930). *Developmental anatomy: A textbook and laboratory manual of embryology.* Philadelphia: Saunders.

Auerbach, K. G., & Danner, S. C. (1988). Measuring sucking patterns [letter]. *Journal of Pediatrics, 112,* 159.

Babcock, C. J. (1938). Feed flavors in milk and milk products. *Journal of Dairy Science, 21,* 661–667.

Barker, E. (1980). *Sensory evaluation of human milk.* Unpublished masters thesis, University of Manitoba, Winnipeg, Canada.

Bassette, R., Fung, D. Y. C., & Mantha, V. R. (1986). Off-flavors in milk. *CRC Critical Reviews in Food Science and Nutrition, 24,* 1–52.

Beauchamp, G. K., & Cowart, B. J. (1990). Preferences for high salt concentrations among children. *Developmental Psychology, 26,* 539–545.

Beauchamp, G. K., Cowart, B. I., Mennella, J. A., & Marsh, R. R. (1994). Infant salt taste: Developmental, methodological and contextual factors. *Developmental Psychobiology, 27,* 353–365.

Beauchamp, G. K., Cowart, B. J., & Moran, M. (1986). Developmental changes in salt acceptability in human infants. *Developmental Psychobiology, 19,* 17–25.

Beauchamp, G. K., & Maller, O. (1977). The development of flavor preferences in humans: A review. In M. R. Kare & O. Maller (Eds.), *The chemical senses and nutrition* (pp. 291–311). New York: Academic Press.

Bilkó, A., Altbacker, V., & Hudson, R. (1994). Transmission of food preference in the rabbit: The means of information transfer. *Physiology & Behavior, 56,* 907–912.

Birch, L. L. (1979). Dimensions of preschool children's food preferences. *Journal of Nutrition Education, 11,* 77–80.

Birch, L. L. (1980). The relationship between children's food preferences and those of their parents. *Journal of Nutrition Education, 12,* 14–18.

Birch, L. L., & Deysher, M. (1986). Caloric compensation and sensory specific satiety: Evidence for self-regulation of food intake by young children. *Appetite, 7,* 323–331.

Blakeslee, A. F. (1932). Genetics of sensory thresholds: Taste for phenyl thio carbamide. *Proceedings of the National Academy of Sciences, USA, 18,* 120–130.

Bossey, J. (1980). Development of olfactory and related structures in staged human embryos. *Anatomy and Embryology (Berlin), 161,* 225–236.

Bradley, R. A. (1757). *A general treatise of agriculture.* London: Johnston.

Bradley, R. M. (1972). Development of the taste bud and gustatory papillae in human fetuses. In J. F. Bosma (Ed.), *The third symposium on oral sensation and perception: The mouth of the infant.* Springfield, IL: Thomas.

Bridger, W. H. (1962). Ethological concepts and human development. *Recent Advances in Biological Psychiatry, 4,* 95–107.

Buck, L., & Axel, R. (1991). A novel multigene family may encode for odorant receptors: A molecular basis for odor recognition. *Cell, 65,* 175–187.

Burt, J. V., & Hertzler, A. A. (1978). Parental influence on the child's food preference. *Journal of Nutrition Education, 10,* 127–130.

Campbell, R. G. (1976). A note on the use of feed flavour to stimulate the feed intake of weaner pigs. *Animal Production, 23,* 417–419.

Capretta, P. J., Petersik, J. T., & Steward, D. J. (1975). Acceptance of novel flavours is increased after early experience of diverse taste. *Nature, 254,* 689–691.

Capretta, P. J., & Rawls, L. H. (1974). Establishment of a flavor preference in rats: Importance of nursing and weaning experience. *Journal of Comparative and Physiological Psychology, 86,* 670–673.

Cernoch, J. M., & Porter, R. H. (1985). Recognition of maternal axillary odors by infants. *Child Development, 56,* 1593–1598.

Cobo, E. (1973). Effect of different doses of ethanol on the milk-ejecting reflex in lactating women. *American Journal of Obstetrics and Gynecology, 115,* 817–821.

Conel, J. L. (1939). *The post-natal development of the human cerebral cortex. I. Cortex of the newborn.* Cambridge, MA: Harvard Univ. Press.

Crook, C. K. (1978). Taste perception in the newborn infant. *Infant Behavior and Development, 1,* 52–69.

Davis, C. M. (1928). Self selection of diet by newly weaned infants: An experimental study. *American Journal of Diseases of Children, 36,* 631–679.

Desor, J. A., Maller, O., & Andrews, K. (1975). Ingestive responses of human newborns to salty, sour and bitter stimuli. *Journal of Comparative and Physiological Psychology, 89,* 966–970.

Desor, J. A., Maller, O., & Turner, R. E. (1977). Preference for sweet in humans: Infants, children and adults. In J. M. Weiffenbach (Ed.), *Taste and development: The genesis of sweet preference.* Washington, DC: US Government Printing Office.

DeSnoo, K. (1937). Das trinkende kind in uterus. *Monatssch fur Geburtshilfe Gynaekol, 105,* 88–97.

Dougherty, R. W., Shipe, W. F., Gudnason, G. V., Ledford, R. A., Peterson, R. D., & Scarpellino, R. (1962). Physiological mechanisms involved in transmitting flavors and odors to milk. I. Contribution of eructated gases to milk flavor. *Journal of Dairy Science, 45,* 472–476.

Doty, R. L. (1986). Ontogeny of human olfactory function. In W. Breipohl (Ed.) *Ontogeny of olfaction: Principles of olfactory maturation in vertebrates.* New York: Springer-Verlag.

Ebers Papyrus (circa 1500 B.C.) *The Papyrus Ebers* [trans. by C. P. Bryan, 1931], (pp. 82–87). New York: Appleton.

Engen, T. (1982). *The perception of odors.* New York: Academic Press.

Engen, T., Lipsitt, L. P., & Kaye, H. (1963). Olfactory responses and adaptation in the human neonate. *Journal of Comparative and Physiological Psychology, 56,* 73–77.

Ettmueller, M. (1699). *Emullerus abrigd'd or A compleat system of the theory and practice of physic.* [trans. anonymous, London.

Fildes, V. (1986). *Breasts, bottles and babies: A history of infant feeding.* Edinburgh: Edinburgh Univ. Press.

Food Advisory Committee Report on the Review of Additives in Foods Specially Prepared for Infants and Children. (1992). FdAC/REP/12. MAFF, HMSO, London.

Galef, B. G. (1971). Social effects in the weaning of domestic rat pups. *Journal of Comparative and Physiological Psychology, 75,* 358–362.

Galef, B. G., & Clark, M. M. (1971). Social factors in the poison avoidance and feeding behavior of wild and domesticated rat pups. *Journal of Comparative and Physiological Psychology, 75,* 341–357.

Galef, B. G., & Clark, M. M. (1972). Mother's milk and adult presence: Two factors determining initial dietary selection by weaning rats. *Journal of Comparative and Physiological Psychology, 78,* 220–225.

Galef, B. G., & Henderson, P. W. (1972). Mother's milk: A determinant of the feeding preferences of weaning rat pups. *Journal of Comparative and Physiological Psychology, 78,* 213–219.

Galef, B. G., & Sherry, D. F. (1973). Mother's milk: A medium for transmission of cues reflecting the flavor of mother's diet. *Journal of Comparative and Physiological Psychology, 83,* 374–378.

Ganchrow, J. R., Steiner, J. E., & Munif, D. (1983). Neonatal facial expressions in response to different qualities and intensities of gustatory stimuli. *Infant Behavior and Development, 6,* 473–484.

Goldman, A. S., Atkinson, S. A., & Hanson, L. Å. (Eds.) (1987). *Human lactation 3: The effects of human milk on the recipient infant.* New York: Plenum.

Guillemeau, J. (1635). *Childbirth or The happy delivery of women . . . To which is added a treatise of the diseases of infants and young children: With the cure of them.* [Trans. anonymous, London]; Printed by Anne Griffin for Joyce Norton and Richard Whitaker.

Gullberg, E. I., Ferrel, F., & Christensen, H. D. (1986). Effects of postnatal caffeine exposure through dam's milk upon weanling rats. *Pharmacology Biochemistry and Behavior, 24,* 1695–1701.

Harris, G., & Booth, D. A. (1987). Infants' preference for salt in food: Its dependence upon recent dietary experience. *Journal of Reproductive and Infant Psychology, 5,* 97–104.

Hauser, G. I., Chitayat, D., Berbs, L., Braver, D., & Mulbauer, B. (1985). Peculiar odors in newborns and maternal pre-natal ingestion of spicy foods. *European Journal of Pediatrics, 44,* 403.

Hepper, P. G. (1987). The amniotic fluid: An important priming role in kin recognition. *Animal Behavior, 35,* 1343–1346.

Hill, D. L., & Mistretta, C. M. (1990). Developmental neurobiology of salt taste sensations. *Trends in Neuroscience, 13,* 188–195.

Hunt, P. S., Kraebel, K. S., Rabine, H., Spear, L. P., & Spear, N. E. (1993). Enhanced ethanol intake in preweanling rats following exposure to ethanol in a nursing context. *Developmental Psychobiology, 26,* 133–153.

Kagan, J. (1971). *Changes and continuity in infancy.* New York: Wiley.

Kajuira, H., Cowart, J., & Beauchamp, G. K. (1992). Early developmental changes in bitter taste responses in human infants. *Developmental Psychobiology, 25,* 375–386.

Korner, A. F., Chuck, B., & Dontchos, S. (1968). Organismic determinants of spontaneous oral behavior in neonates. *Child Development, 39,* 1145–1157.

Krebs, R. (1953). *Making friends is our business—100 years of Anheuser-Busch.* St. Louis, Missouri: A-B.

Leon, M. (1992). The neurobiology of filial learning. *Annual Review of Psychology, 43,* 77–98.

Liley, A. W. (1972). Disorders of amniotic fluid. In N. S. Assali (Ed.), *Pathophysiology of gestation: Fetal placental disorders, Vol. 2.* New York: Academic Press.

Little, R. E., Anderson, K. W., Ervin, C. H., Worthington-Roberts, B., & Clarren, S. K. (1989). Maternal alcohol use during breast feeding and infant mental and motor development at one year. *New England Journal of Medicine, 321,* 425–430.

London, R. M., Snowdon, C. T., & Smithana, J. M. (1979). Early experience with sour and bitter solutions increases subsequent ingestion. *Physiology & Behavior, 22,* 1149–1155.

Mace, J. W., Goodman, S. I., Centerwall, W. R., & Chinnock, R. F. (1976). The child with an unusual odor: A clinical resumé. *Clinical Pediatrics, 15,* 57–62.

MacFarlane, A. J. (1975). Olfaction in the development of social preferences in the human neonate. *Ciba Foundation Symposium, 33,* 103–117.

Mainardi, M., Poli, M., & Valsecchi, P. (1989). Ontogeny of dietary selection in weaning mice: Effects of early experience and mother's milk. *Biology of Behavior, 14,* 185–194.

Maone, T. R., Mattes, R. D., Bernbaum, J. C., & Beauchamp, G. K. (1990). A new method for delivering a taste without fluids to preterm and term infants. *Developmental Psychobiology, 23,* 179–191.

McDaniel, M. R. (1980). Off-flavors in human milk. In G. Charalambous (Ed.), *The analysis and control of less desirable flavors in foods and beverages* (pp. 267–291). New York: Academic Press.

Mennella, J. A. (1995). Mother's milk: A medium for early flavor experiences. *Journal of Human Lactation, 11,* 39–45.

Mennella, J. A., & Beauchamp, G. K. (1991a). Maternal diet alters the sensory qualities of human milk and the nursling's behavior. *Pediatrics, 88,* 737–744.

Mennella, J. A., & Beauchamp, G. K. (1991b). The transfer of alcohol to human milk: Effects on flavor and the infant's behavior. *New England Journal of Medicine, 325,* 981–985.

Mennella, J. A., & Beauchamp, G. K. (1993a). Beer, breast feeding and folklore. *Developmental Psychobiology, 26,* 459–466.

Mennella, J. A., & Beauchamp, G. K. (1993b). The effects of repeated exposure to garlic-flavored milk on the nursling's behavior. *Pediatric Research, 34,* 805–808.

Mennella, J. A., & Beauchamp, G. K. (1994). The infant's responses to flavored milk. *Infant Behavior and Development, 17,* 819.

Mennella, J. A., Johnson, A., & Beauchamp, G. K. (1995). Garlic ingestion by pregnant women alters the odor of amniotic fluid. *Chemical Senses, 20,* 207–209.

Mennella, J. A., & Beauchamp, G. K. (1996a). The human infants' responses to vanilla flavors in human milk and formula. *Infant Behavior and Development, 19,* 13–19.

Mennella, J. A., & Beauchamp, G. K. (1996b). Developmental changes in the infants' acceptance of protein-hydrolysate formula and its relation to mothers' eating habits. *Journal of Developmental and Behavioral Pediatrics, 17,* 386–391.

Mixsell, H. R. (1916). A short history of infant feeding. *Archives of Pediatrics, 33,* 282–293.

Morrill, J. L., & Dayton, A. D. (1978). Effect of feed flavor in milk and calf starter on feed consumption and growth. *Journal of Dairy Science, 61,* 229–232.

Nakashima, T., Kimmelman, C. P., & Snow, J. B. (1985). Immunohistopathology of human olfactory epithelium, nerve and bulb. *Laryngoscope, 95,* 391–396.

Nolte, D. L., & Provenza, F. D. (1991). Food preferences in lambs after exposure to flavors in milk. *Applied Animal Behavior Science, 32,* 381–389.

Nolte, D. L., Provenza, F. D., & Balph, D. F. (1990). The establishment and persistence of food preferences in lambs exposed to selected foods. *Journal of Animal Science, 68,* 998–1002.

Pangborn, R. M., Guinard, J.-X., & Davis, R. G. (1988). Regional aroma preferences. *Food Quality and Preference, 1,* 11–19.

Peterson, R., & Rainey, L. H. (1910–1911). The beginnings of mind in the newborn. *Bulletin of the Lying-in Hospital of the City of New York, 7,* 99–122.

Phillips, D. S., & Stainbrook, G. L. (1976). Effects of early alcohol exposure upon adult learning abilities and taste preferences. *Physiological Psychology, 4,* 473–475.

Pliner, P., & Pelchat, M. L. (1986). Similarities in food preferences between children and their siblings and parents. *Appetite, 7,* 333–342.

Postman, N. (1982). *The disappearance of childhood.* New York: Dell.

Pritchard, J. A. (1965). Deglutition by normal and anencephalic fetuses. *Obstetrics and Gynecology, 25,* 289–297.

Provenza, F. D., & Balph, D. F. (1987). Diet learning by domestic ruminants: Theory, evidence and practical implications. *Applied Animal Behavior Science, 18,* 211–232.

Ray, J. W., & Klesges, R. C. (1993). Influences on the eating behavior of children. Prevention and treatment of children. *Annals of the New York Academy of Sciences, 699,* 57–69.

Reiser, J., Yonas, A., & Wilkner, K. (1976). Radial localization of odors by human newborns. *Child Development, 47,* 856–859.

Rolls, B. J., Rowe, E. S., & Rolls, E. T. (1982). How sensory properties of foods affect human feeding behavior. *Physiology & Behavior, 29,* 407–417.

Robinovitch, L. G. (1903). Infantile alcoholism. *Quarterly Journal of Inebriety, 25,* 231–236.

Rosenstein, D., & Oster, H. (1990). Differential facial responses to four basic tastes in newborns. *Child Development, 59,* 1555–1568.

Routh, C. H. F. (1879). *Infant feeding and its influence on life.* New York: William Wood, and Company.

Rozin, P. (1976). The selection of food by rats, humans and other animals. In J. Rosenblatt, R. A. Hinde, C. Beer, & E. Shaw (Eds.), *Advances in the study of behaviors: Vol. 6.* (pp. 21–76). New York: Academic Press.

Rozin, P. (1980). Human food selection: Why do we know so little and what can we do about it? *International Journal of Obesity, 4,* 333–337.

Rozin, P. (1982). "Taste-smell confusions" and the duality of the olfactory sense. *Perception and Psychophysics, 31,* 397–401.

Rozin, P. (1984). The acquisition of food habits and preferences. In J. D. Mattarazzo, S. M. Weiss, J. A. Herd, N. E. Miller, & S. M. Weiss (Eds.), *Behavioral health: A handbook of health enhancement and disease prevention* (pp. 590–607). New York: Wiley.

Sarnat, H. B. (1978). Olfactory reflexes in the newborn infant. *Journal of Pediatrics, 92,* 624–626.

Schaal, B. (1986). Presumed olfactory exchanges between mother and neonate in humans. In J. Le Camus, & J. Conier (Eds.), *Ethology and psychology* (pp. 101–110). Toulouse: Privat IEC.

Schaal, B. (1988). Olfaction in infants and children: Development and functional perspectives. *Chemical Senses, 13,* 145–190.

Schaal, B., Orgeur, P., & Porter, R. H. (1994). Short-term flavor preference induced by dietary flavors transferred into mother's milk. *Infant Behavior and Development, 17,* 927.

Schaffer, J. P. (1910). The lateral wall of the cavum nasi in man with special reference to the various developmental stages. *Journal of Morphology, 21,* 613–617.

Self, P. A. F., Horowits, F. D., & Paden, L. Y. (1972). Olfaction in newborn infants. *Developmental Psychology, 7,* 349–363.

Shipe, W. F. (1980). Analysis and control of milk flavors. In G. Charalambous (Ed.), *The analysis and control of less desirable flavors in foods and beverages.* New York: Academic Press.

Shipe, W. F., Bassette, R., Deane, D. D., Dunkley, W. L., Hammond, E. G., Harper, W. J., Kleyn, D. H., Morgan, M. E., Nelson, J. H., & Scanlan, R. A. (1978). Off-flavors of milk: Nomenclature, standards and bibliography. *Journal of Dairy Science, 61,* 855–868.

Shipe, W. F., Ledford, R. A., Peterson, R. D., Scanlan, R. A., Geerken, H. F., Dougherty, R. W., & Morgan, M. E. (1962). Physiological mechanisms involved in transmitting flavors and odors to milk. II. Transmission of some flavor components of silage. *Journal of Dairy Science, 45,* 477–480.

Snyder, L. H. (1931). Inherited taste deficiency. *Science, 74,* 151–152.

Soranus (1st/2nd century A.D.). *Soranus' gynecology,* [trans. by O. Temkin, N. J. Eastman, L. Edelstein, & A. F. Guttmacher, 1956] (pp. 90–103). Baltimore: The Johns Hopkins Univ. Press.

Spielman, A. I., Huque, T., Nagai, H., Whitney, G., & Brand, J. G. (1994). Generation of inositol phosphates in bitter taste transduction. *Physiology & Behavior, 56,* 1149–1155.

Steiner, J. E. (1977). Facial expressions of the neonate infant indication the hedonics of food-related chemical stimuli. In J. M. Weiffenbach (Ed.). *Taste and development: The genesis of sweet preference.* Washington, DC: US Government Printing Office.

Steiner, J. E. (1987). What the neonate can tell us about umami. In Y. Kawamura, & M. R. Kare (Eds.), *Umami: A basic taste.* New York: Dekker.

Tatzer, E., Schubert, M. T., Timischl, W., & Simbruner, G. (1985). Discrimination of taste and preference for sweet in premature babies. *Early Human Development, 12,* 23–30.

Terry, L. M., & Johanson, I. B. (1987). Olfactory influences on the ingestive behavior of infant rats. *Developmental Psychobiology, 20,* 313–332.

Thorhalldottir, A. G., Provenza, F. D., & Balph, D. F. (1990). Ability of lambs to learn about novel foods while observing or participating with social models. *Applied Animal Behaviour Science, 25,* 25–33.

Valsecchi, P., Moles, A., & Mainardi, M. (1993). Does mother's diet affect food selection of weanling wild mice? *Animal Behavior, 46,* 827–828.

Varendi, H., Porter, R. H., & Winberg, J. (1994). Does the newborn baby find the nipple by smell? *Lancet, 334,* 989–990.

Vasquez, M., Pearson, P. B., & Beauchamp, G. K. (1982). Flavor preferences in malnourished Mexican infants. *Physiology & Behavior, 28,* 513–519.

Vilaró, S., Viñas, O., Remesar, X., & Herrera, E. (1987). Effects of chronic ethanol consumption on lactational performance in the rat: Mammary gland and milk composition and pups' growth and metabolism. *Pharmacology Biochemistry and Behavior, 27,* 333–339.

Wickes, I. G. (1953). A history of infant feeding. *Archives of Diseases in Childhood, 28,* 151–158.

Wilson, D. A., & Sullivan, R. M. (1994). Neurobiology of associative learning in the neonate: Early olfactory learning. *Behavioral and Neural Biology, 61,* 1–18.

Wuensch, K. L. (1978). Exposure to onion taste in mother's milk leads to enhanced preference for onion diet among weanling rats. *Journal of General Psychology, 99,* 163–167.

Wysocki, C. J., & Beauchamp, G. K. (1984). The ability to smell androstenone is genetically determined. *Proceedings of the National Academy of Sciences, USA, 86,* 7976–7978.

Index